T0326347

Portable Microwave and mmWave Radars for Contactless Healthcare

Published 2025 by River Publishers

River Publishers

Alsbjergvej 10, 9260 Gistrup, Denmark

www.riverpublishers.com

Distributed exclusively by Routledge

605 Third Avenue, New York, NY 10017, USA

4 Park Square, Milton Park, Abingdon, Oxon OX14 4RN

Portable Microwave and mmWave Radars for Contactless Healthcare / by Emanuele Cardillo, Changzhi Li.

Routledge is an imprint of the Taylor & Francis Group, an informa business

ISBN 978-87-7004-752-4 (paperback)

ISBN 978-87-7004-754-8 (online)

ISBN 978-87-7004-753-1 (ebook master)

DOI 10.1201/9781003609483

A Publication in the River Publishers Series in Rapids

Portable Microwave and mmWave Radars for Contactless Healthcare

Emanuele Cardillo

Department of Engineering, University of Messina, Messina, Italy

Changzhi Li

Department of Electrical and Computer Engineering,
Texas Tech University, Lubbock, TX, USA

NEW YORK AND LONDON

Contents

Preface

For many years, the scientific community has put significant effort into the development of ever more advanced microwave and mmWave radar systems for healthcare applications. The main reasons behind the success of this technology, i.e. alleviated privacy concerns, compactness, immunity to different light and temperature conditions, the contactless nature of the detection and the great number of accurate and advance information provided to the users, emerged.

As a consequence, different books have reviewed and described the latest advancements in the field. However, young students, researchers and professionals need a reference book which clearly and concisely describes the main principles of portable radars and identifies the relevant applications of this technology; this is exactly the aim of this book, which aspires to serve as both an entry point for students and researchers new to the topic and a reference for those seeking to explore specific aspects of radar technologies in healthcare.

About the Authors

Emanuele Cardillo received his M.Sc. degree in Electronic Engineering from the University of Messina, Italy, and Ph.D degree from the University Mediterranea of Reggio Calabria, Italy, in 2013 and 2018, respectively. He is currently Senior Assistant Professor and the head of the Microwave Electronics Laboratory at the University of Messina. His research interests are focused on the microwave and mm-wave electronics field, mainly on portable radars, microwave measurements, circuit design and realization. He is the recipient of the Italian 2022 MUR PRIN project: "Contactless And ReliAble Movement invEstigation with miLLimeter-waves rAdars". He was the recipient of the 2018 IEEE Sensors Journal Best Student Paper Award and the 2018 IEEE MTT-S award from the IEEE Sensors Council Italy Chapter and the IEEE MTT-S/AP-S Italy Chapter, respectively. He is the chair of the TC "Microwave and Millimeter-Wave Radar Sensors" of the IEEE Sensors Council Italy Chapter and TC Member of the IEEE MTT-28: Biological Effects and Medical Applications of RF and Microwaves.

Changzhi Li received his Ph.D. degree in electrical engineering from the University of Florida, Gainesville, FL, in 2009. He is a Professor at Texas Tech University. His research interests are microwave/millimeter-wave sensing for healthcare, security, energy efficiency, structural monitoring, and human-machine interface. Dr. Li was an IEEE Microwave Theory and Techniques Society (MTT-S) Distinguished Microwave Lecturer, in the Tatsuo Itoh class of 2022â2024. He was a recipient of the IEEE MTT-S Outstanding Young Engineer Award, the IEEE Sensors Council Early Career Technical Achievement Award, the ASEE Frederick Emmons Terman Award, the IEEE-HKN Outstanding Young Professional Award, and the NSF Faculty Early CAREER Award. He is the General Chair of the 2024 IEEE Radio Wireless Week (RWW) in San Antonio, TX. Dr. Li is a Fellow of the IEEE and the National Academy of Inventors (NAI).

List of Notations and Abbreviations

RADAR	Radio detection and ranging
ADAS	Advanced driver assistance systems
mawWave	Millimeter wave
RF	Radio frequency
MIMO	Multiple-input multiple-output
ECG	Electrocardiogram
IR	Infrared radiation
PIR	Passive infrared
IC	Integrated circuit
GPS	Global positioning systems
RFID	Radio frequency identification
PCB	Printed circuit board
HF	High frequency
VHF	Very high frequency
UHF	Ultra high frequency
IEEE	Institute of Electrical and Electronics Engineers
RRE	Radar range equation
SNR	Signal to noise ratio
EM	Electromagnetic
RCS	Radar cross section
PDF	Probability density functions
ROC	Receiver operating characteristic
CA-CFAR	Cell-average constant false alarm rate
LOS	Line of sight
CW	Continuous wave
FFT	Fast Fourier transform
FMCW	Frequency modulated continuous wave
ADC	Analog-to-digital converter
LADAR	Laser detection and ranging
LO	Local oscillator

STFT	Short-time Fourier transform
AOA	Angle-of-arrival
SISO	Single-input single-output
AWGN	Additive white Gaussian noise
EQV	Field of view
ULA	Uniform linear array
URA	Uniform rectangular array phase modulation
TDM	Time division multiplexing
FDM	Frequency division multiplexing
ETA	Electronics travel aid
VNA	Vector network analyzer
RGB-D	Red green blue-depth
IR-UWB	Impulse radio-ultra wide band
DACM	Differentiate and cross-multiply

Portable Microwave and mmWave Radars for Contactless Healthcare

1.1 Introduction

"Sending electromagnetic waves to obtain information about the surrounding space", this is the essence of the radar operations developed over the last 150 years. The acronym for radio detection and ranging properly describes the main operations of early radars but underestimates the characteristics and potential of modern radars.

By exploiting the capability to detect the presence and range of objects, namely targets, earlier versions of the radar were mainly exploited for military applications. Military applications still benefit from the features of radars, from tactical radars to control the airspace in the search for aircraft and missiles, to monitor the Earth's surface in search for moving ground vehicles, and look after special sites such as Earth resources, cultural sites, and military facilities [1].

Moreover, radars are successfully exploited for civilian applications, e.g., police radars to measure the speed of vehicles, Earth monitoring radars to study the environmental characteristics of the Earth, e.g., in terms of water levels and ice thickness, weather radars, often considered to be between the military and civilian fields to detect precipitations, space observation and home security systems.

The automotive sector gave a great push to the field; indeed, radars are used both as inner and outer ADAS sensors in many cars for in-cabin monitoring applications and to avoid collisions, thus making the driving safer [2], [3].

This allowed the price of radars to decrease and to provide increasingly better devices as a consequence of both academic and industry effort.

Of course, it is not possible to comprehensively list all the possible radar applications. This assertion is strengthened by the continuously growing number of new applications of modern radars throughout different fields and particularly throughout the biomedical sector.

Some insight about the main applications of portable radars in healthcare, the current health monitoring technologies and a comparison between the different technologies for contactless monitoring will be given in the next sections, whereas more detailed examples will be provided in Section 1.3.

1.1.1 History of Radar

The first appearance of radar systems is commonly ascribed to World War II, driven by military needs. However, the first discoveries related to radar theory are found in the classic electromagnetic theory developed by James Clerk Maxwell in the 1860s, and the first demonstration of transmission and reflection of radio waves by Heinrich Hertz in the 1880s. During World War II, for the first time Great Britain deployed a great number of radars along its coasts to detect enemy targets, the so-called Chain Home. On the other hand, it is very interesting that the Germanic counterpart tried to use the British Chain Home to their advantage. They deployed a passive radar system called the Klein Heidelberg Parasit or Heidelberg-Gerät, at seven sites, using the British Chain Home radars as non-cooperative transmitters. Passive radars detect obstacles from the echoes received from non-cooperative transmitters. In this case, the British Chain Home is the non-cooperative transmitter whereas a modern example is the home Wi-Fi signal serving as the transmitter for a passive receiver to sense indoor human activities.

The timeline of the events leading to the radar development is given in Table 1.1 [4], [5].

Clearly, from the earlier version of RF detection systems to the radar systems employed during World War II, a lot of effort has been expended. Among the many applications, e.g., in the industrial, military, space and automotive fields, the next sections will introduce the main features of portable radars in the healthcare field.

Table 1.1: Timeline of radar development.

Year	Main event
1860s	James Clerk Maxwell developed the classic electromagnetic theory
1880s	Heinrich Hertz demonstrated the transmission and reflection of radio waves
1900s	Christian Hülsmeyer and Nikola Tesla detected moving ships
1920s	Robert Alexander Watson Watt detected storms through a radar system named Huff Duff
1922	Albert Hoyt Taylor and Leo Clifford Young demonstrated ship detection based on an interference due to a ship passing
1924	Edward Victor Appleton proved the existence of, and measured the range of, the ionosphere through the first frequency-modulated continuous-wave radar
1930	Lawrence A. "Pat" Hyland detected an aircraft by radar, leading to one of the first radar patents in 1934
1935	Robert Alexander Watson Watt patented the first pulsed radar system for detecting enemy aircraft. In his experiment, he detected a Handley Page Heyford bomber using a passive radar, whereby the non-cooperative transmitter was the BBC shortwave transmitter at Daventry
1937	First ship equipped with ship-tracking radar designed 17 years before by A. Hoyt Taylor
1939	Great Britain deployed a chain of radars along its coasts to detect enemy targets

1.1.2 Main Healthcare Applications of Portable Radars

The great success of radar sensing systems goes beyond the capability to detect range and speed. Indeed, modern radars are able to measure very small displacements, even to beneath the impressive threshold of one micrometer, and detect the different Doppler components of a moving object [6], [7].

These facts suggest that the non-contact sensing of life activities for healthcare can be synthetically divided into two main fronts, one related to vital sign detection, the other related to human activity analysis. This is of course a very synthetic and in part reductive classification of the topic, but it well identifies two directions undertaken by the scientific community, one based on phase analysis in term of displacement and Doppler detection, as if the target was a single point or a parallel plane, the other one based on the micro-Doppler signature, whereby the target is treated as a complex rigid or non-rigid body with each sub-component moving independently.

The detailed vital sign detection principles, mainly based on the displacement measurement induced by tiny physiological movements, in turn gathered from the phase analysis of the received echo, will be detailed in Section 2.2.

Vital sign detection has been successfully exploited to monitor infants' or adults' sleeping, and to detect abnormal health conditions and relevant cardiovascular metrics as the blood pressure waveform [8]–[12]. If an abnormal behavior is detected within a pre-established time window, the user or an assisting person can be alerted.

However, the task to detect vital signs should not be seen only with the purpose to accurately measure breathing signals and heartbeat, but it is often exploited as the key sign of human presence [13]. As a consequence, it leads to many potential applications such as searching for people after a natural disaster, continuous authentication and in-cabin detection [2], [14]–[18]. Moreover, occupancy sensing often relies on the detection of vital signs, but the success of the subject detection can be improved by searching for possible signs of the normal movement of the body due to daily activities [19]–[21].

Almost 50 years have passed since the first demonstration of vital sign detection [22], while researchers' effort has been focused on improving the accuracy, reducing the power consumption, weight and size of radars with cost-effective technologies [23].

On the other hand, activity/movement analysis can be separated from the vital sign topic because it is based on a different concept. Although almost everything concerning radar detection is based on phase analysis, directly or indirectly, of the target as a single punctiform object, the activity detection relies mainly on the micro-Doppler behavior of the subject [24].

The scientific literature reports on many contributions related to human activity recognition; this topic has been investigated both to gather human biomechanical parameters, useful to monitor the health status of the body and to assist rehabilitation procedures, to distinguish falls from normal activities, particularly for elderly subjects, and to sense and classify human gestures for human–machine interaction [2], [3], [25]–[30].

In addition to the aforementioned most famous topics, the scientific community tried to exploit radar features for different purposes related to the healthcare field. Many studies report on radar systems exploited as mobility aids for blind and visually impaired people, whereby the radar is the sensing element able to detect potential obstacles through the path of the subject and communicate the real time distance and nature of the target, i.e. if it is human or non-human [31]–[35].

Portable radars have been successfully employed as tools to facilitate the communication of subjects affected by neurodegenerative disorders, in detail, by detecting intentional head movements or eye blinking to be translated into commands or messages [36]–[40]. The scientific literature also shows different examples related to the radar detection of inattentive driver behavior based on head movements and body posture [41]–[43].

MIMO radars are also exploited for their capability to detect the angle-of-arrival of the echo, thus enabling the selective detection of vital signs and motion parameters at different angular positions [44]–[47]. This is usually performed by using more receivers and taking advantage of the phase difference due to the different round-trip distance between the target and each different receiver. Moreover, multiple transmitters can be used to increase the resolution with great scientific effort put into investigating different antenna configurations [44], [48]. MIMO radars have been widely used for biomedical applications, including real-time heart rate estimation across multiple subjects [44].

1.1.3 Current Health Monitoring Technologies

Many different technologies have been proposed in the scientific literature for healthcare applications. It can be stated that the perfect sensor does not exist and, in many cases, the best solution is a trade-off between different features of different technologies and the best results can be achieved by using multiple technologies simultaneously.

This section systematically describes the current health monitoring technologies that can be complementary to radar to help the reader select the best sensor for a specific application.

A first categorization can be made in terms of wearable and contactless health monitoring sensors. Wearable sensors, due to the contact with the user, show good performance in terms of reliability, particularly in the presence of random body motion [49]–[53]. Examples of wearable health monitoring devices are respiratory belts, smartwatches, and pulse oximeters, which might result in uncomfortable user experiences due to the obvious contact with the body. Although these systems are well-accepted by people, particularly sportsman, some subjects, including unhealthy patients, could refuse to wear contact-based sensors, thus severely restricting their application field [44]. An additional limitation is that, despite the fact that they can simultaneously measure more than one physiological parameter, they cannot usually monitor multiple subjects. They are mainly exploited for measuring cardiopulmonary functions [45]; to monitor the heart parameters over time, ECG represents the gold standard, and it is indeed used both in clinics and hospitals. In detail the electrical signal involved in heart muscle contraction is measured by electrodes attached to different parts of the body. Other contact devices measure the variation of blood volume at the skin with an optical sensor.

On the other hand, the measurement of breathing activity is not likewise standardized, although it is a key indicator of health status. Despite great advancement in the available technology, the breathing rate is measured manually, particularly in hospitals, with an operator who counts the number of breaths following a visual assessment. There are three available options to measure respiration, according to [4], [54]: measurement of oxygen saturation, airflow and respiratory movement.

Pulse oximetry relies on the presence of a pulsed signal generated by arterial blood, and the fact that hemoglobin (Hb) has different absorption spectra [55]. Hemoglobin absorbs an amount of light which varies depending on the oxygen saturation. By analyzing the output of pulse oximetry, it is possible to gather if a breathing disease is present, but without any information concerning the respiratory rate.

A spirometer is used to measure the airflow, volume and breathing rate by exploiting a pressure transducer in a face mask. However, spirometers are considered uncomfortable due to the face mask requirement and the airflow resistance and can affect the normal respiratory activity.

The chest motion due to the body volume change resulting from the respiratory activity can be measured indirectly by exploiting impedance plethysmographs, strain gauge measurement of thoracic circumference and pneumatic respiration transducers [4]. The plethysmograph requires the patient to be stationary whereas the performance of belts and electrodes for respiration monitoring significantly degrades over time.

Great effort has been devoted to detecting physiological parameters without any contact. The most employed technologies are based on thermal imaging and optical systems [44].

Among the different technologies, camera-based systems are considered very effective to measure movement parameters. As an example, stereophotogrammetry is used to analyze the human motion parameters to better understand musculo-skeletal system physiopathology [56], [57]. Cameras are also used to assess the human presence, taking the role of occupancy sensors.

The acquired images are analyzed to highlight the human presence but usually requires computationally costly signal processing algorithms, and their performance are affected by different light conditions. The sensitivity to different light conditions is also a serious limiting factor when cameras are employed as photoplethysmography systems, i.e. to measure the changes in skin color due to physiological activity [58], [59].

Moreover, whereas cameras are effective in detecting the target angular position, they require a more complex procedure to accurately detect the range of the target. Finally, they might raise privacy concerns, thus limiting the possible applications. Stereophotogrammetry systems often exploit infrared cameras and require the user to wear special markers to separate the different parts of the body [56]. This is of course another limiting factor for the users' comfort.

Ultrasonic sensors transmit ultrasonic signals whose wavelength is approximately 1.9 cm, and can be used as occupancy sensors whose principle is based on the time-of-flight measurement. They can detect stationary and moving objects and measure their speed based on Doppler shift but usually only in a short range, which is even more limited when smooth surfaces need to be detected, and even worse if the angle of incidence of the ultrasonic beam is smaller. Moreover, ultrasonic devices exhibit a relatively large radiation beam that cannot be easily controlled and are sensitive to environmental conditions. On the other hand, it is a cost-effective and widely available technology [32].

The scientific literature also shows examples of thermal imaging extraction of vital signs based on the infrared radiation emitted by the body [60], [61]. Depending on the analyzed part of the body, different outcomes can be obtained. As an example, breathing activity is related to the temperature gradient caused by breathing out, whereas heart activity is due to the temperature change due to blood flow inside superficial arteries [44]. The main limitations are due to random body motion artifacts and to the high cost of the sensors.

1.1.4 Pros and Cons of Radars for Contactless Monitoring

Radars can be considered a valid alternative to other modalities, and they overcome the performance of the competitors in many occurrences.

Being contactless sensors, compared to wearable sensors, radars do not affect the user comfort and do not modify the parameters being measured, thus representing the ideal technology for continuous and long-term monitoring applications.

They are immune to the different environmental and light conditions, a trivial exception is represented by the dependence on the target material that, by changing the target radar cross section, in turn affects the maximum range and thus the range accuracy.

Compared to ultrasonic and optical systems, radars present the following advantages:

- Compact size and lightweight: the higher the frequency, the smaller the dimension; in radars, most of the total occupied space is due to the antenna whose dimensions are strongly related to the wavelength. Therefore, increasing the frequency reduces the wavelength and in turn the size. In Figure ??, two radars working at different operating frequencies are shown, whereby although different substrates have been used, the scaling size of the antennas can be observed.
- Immunity to ambient light conditions: as opposed to optical and ultrasonic systems, radars are immune to light and environmental conditions.
- High theoretical resolutions: the range resolution of FMCW radars improves with a higher bandwidth whereas the speed resolution improves with a longer active measurement interval. These aspects will be described in Section 2.2. It is worth noting that this is particularly true for state-of-the-art mmWave radar sensors with a large bandwidth. As a matter of fact, modern semiconductor technology has made it possible to achieve such a large bandwidth with low-cost portable radar chips and systems [62].
- Customizable field-of-view: the radar field of action greatly depends on the antenna radiation pattern which can be tailored by modifying the design according to the specific needs.
- Alleviated privacy concerns: radars do not acquire images thus it is in general very challenging recognizing a specific individual.
- Able to work both with stationary and moving targets: other technologies employed for occupancy sensing work mainly for moving targets. As an example, PIR sensors are able to measure temperature variations due to a moving subject, whereas FMCW radars can also measure stationary subjects.
- Advanced information: radars are able to provide advanced information, as very small displacements related to the physiological activity, and extract the micro-Doppler signature where the different velocity components of the different parts of the body can be recognized. Even though other sensing modalities may also claim advanced information, radars can easily detect Doppler and range information with low computational load, since obtaining this information is a matter of measuring the frequency components of a signal.

Potentially, radar can be in the same price range or even cheaper than ultrasonic and infrared systems, since radar can be completely built on silicon/semiconductor chips, while the other two technologies need extra processes to fabricate them.

The user should select the best technology for a specific application by considering that an ultrasonic signal propagates well in solid media but has high loss in air, whereas radar has low loss in air but encounters high loss in water and some other materials.

Figure 1.1: Radar boards working at the central frequency of (a) 5.8 GHz, (b) 24 GHz. The patch sizes are approximately (a) 17 mm × 11 mm, and (b) 6 mm × 4 mm, [J. -M. Muñoz-Ferreras, Z. Peng, Y. Tang, R. Gómez-García, D. Liang and C. Li, "Short-Range Doppler-Radar Signatures from Industrial Wind Turbines: Theory, Simulations, and Measurements," in IEEE Transactions on Instrumentation and Measurement, vol. 65, no. 9, pp. 2108-2119, Sept. 2016, doi: 10.1109/TIM.2016.2573058].

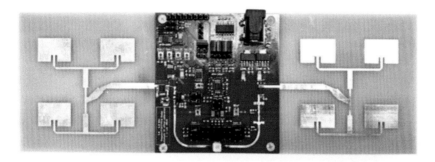

(a)

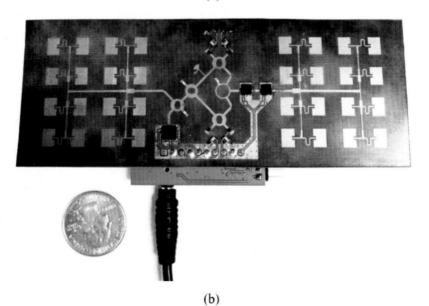

(b)

Moreover, some effort should be devoted to making users willing to accept an electromagnetic based system. Indeed, although the transmitted power levels are limited by the regulations, likewise other sources of electromagnetic waves, often people are reluctant to use new and somehow mysterious technologies.

Radar performance is also limited by some issues that can be considered disadvantages compared to other technologies because they also affect other systems. As an example, since radars measure radial quantities, they are in general not able to separate self-motion from the target motion [65]–[67]. Moreover, the target random body motion is considered a great challenge, particularly when the goal is measuring vital signs, whose extent can be tiny compared to the body motion [68]–[73].

Finally, the presence of hardware level imperfections, excessive clutter, i.e. unwanted targets (both non-human and human additional targets) in the scenario, can worsen the overall performance.

1.1.5 The Role of Microwave and mmWave Electronics

RF, microwave and mmWave electronics are widespread and pervasive throughout almost every day applications, e.g. telecommunications, anti-collision systems for vehicles, direct broadcast satellites, GPS, RFID, remote sensing systems for the environment, defense systems and of course radars. By exploiting different operating frequencies, i.e. different operating wavelengths, the electronic circuits can be tailored to a specific application, e.g., long range, large bandwidth, etc.

The RF frequency bandwidth ranges from 30 MHz to 3 GHz, the microwave bandwidth ranges from 3 GHz to 300 GHz, between radio waves and infrared waves, whereas signals whose wavelength is in the order of millimeters, conventionally from 30 GHz to 300 GHz are referred to as mmWaves [74].

Even though the higher frequencies make the analysis and design of microwave devices and systems complicated, microwave electronics shows the following practical advantages which are exploited in many short-range applications including radars:

- The gain of the antennas is proportional to its electrical size. Since the higher the frequency the smaller the wavelength, at higher frequencies for the same physical size the antenna looks electrically bigger thus increasing the gain.

- At higher frequencies, more bandwidth is provided, thus more data can be transmitted by telecommunication systems and better performance can be obtained in radar applications.
- At higher frequencies, i.e. smaller wavelength, the target radar cross section is proportional to the electrical size of the target, which is a beneficial aspect.
- Higher frequencies allow managing a crowded spectrum since at higher frequencies fewer applications make the available spectrum wider.
- As described in the previous section, among the advantages of radars compared to competitors, the higher the frequency, the smaller the dimension, thus more compact and lightweight systems can be designed.
- Microwave electronics is suitable to be designed on planar technologies, as microstrip lines, which allow an easier integration with standard PCB technologies.
- Microwave energy is reflected by metallic materials but penetrates other kinds of materials as some plastics or glass. This is really appreciated to protect or hide the radar behind a cover. The typical behavior of microwaves through different materials is reported in Table 1.2.

Table 1.2: Microwave typical behavior through different materials.

Material	Microwave behavior
Metals	High reflection
Waters	High absorption
Chemical foams	Low attenuation
Dry clothing	Low attenuation
Wet clothing	Up to 20 dB losses
Rain	Up to 6 dB losses
Plastics	0.5 dB to 3 dB losses
Dry wood	Low losses
Wet wood	Up to 20 dB losses
Ice	Up to 10 dB losses

The electromagnetic spectrum is divided into sub-bands depending on the frequency/wavelength. Mnemonic letters are associated with each band to allow professional reference to a specific bandwidth easier, according to the IEEE classification reported in Table 1.3.

Table 1.3: RF/microwave spectrum sub-bands.

IEEE sub-band designation	Sub-band frequency range
Medium frequency	300 kHz–3 MHz
HF	3 MHz–30 MHz
VHF	30 MHz–300 MHz
UHF	300 MHz–1 GHz
L band	1–2 GHz
S band	2–4 GHz
C band	4–8 GHz
X band	8–12 GHz
Ku band	12–18 GHz
K band	18–27 GHz
Ka band	27–40 GHz
V band	40–75 GHz
W band	75–110 GHz
F band	90–140 GHz

Among the different sub-bands, the L band is mainly exploited for navigation purposes, mobile phones, and in military applications, the S band in navigation beacons and wireless networks, the C band for long-distance telecommunications due to good cloud, dust, and rain penetration, the X band for radar and space communications, the Ku band referring to K-under for satellite communications, the K band and Ka band, referring to K-above, for satellite communications, astronomical observations and radar, the V band for high capacity terrestrial mmWave communication and radar. Automotive radars currently work in the K and W bands, roughly around 24 GHz and 77 GHz, approximately, whereas industrial radars work roughly at 60 GHz, within the V band.

2

Radar Operating Principles

In this chapter the main radar operating principles are illustrated. The purpose of the chapter is to explain the most relevant concepts by using a simple and concise language.

2.1 Radar Equation

Radar performance dramatically depends on the level of the received echo. However, at the receiver input, in addition to the signal of interest, noise is received likewise. To properly detect the signal, the signal must overcome the noise level by a certain margin, commonly referred to as signal-to-noise ratio.

Radar system designers use an equation to compute the SNR at the receiver, namely the radar range equation. The RRE is a relatively simple equation which takes into account all the parameters affecting the power budget. The RRE can also be exploited to determine the maximum range at which a particular radar can detect a target, but it can also serve as a means for understanding the factors affecting radar performance.

By considering the scenario depicted in Figure 2.1, where a radar transmitter sends an EM wave which travels through the channel and it is reflected by a target and received by the radar receiver, it is possible to compute the power at the target P_{targ} as:

$$P_{targ} = \frac{P_t G_t A_{fs}}{L_t} \tag{2.1}$$

Figure 2.1: Schematic representation of the typical radar environment.

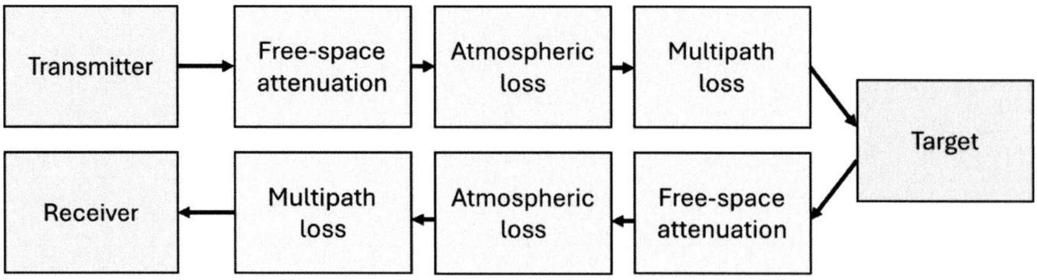

where

- P_t is the transmitted power
- G_t is the transmitting antenna gain
- A_{fs} is the free space attenuation
- L_t is the sum of atmospheric loss and multipath loss, i.e. the propagation of a wave from one point to another by more than one path.

The free space attenuation A_{fs} can be in turn computed as:

$$A_{fs} = \frac{\lambda^2}{(4\pi R)^2} \tag{2.2}$$

where

- R is the target range (line-of-sight distance)
- λ is the operating wavelength.

It is worth noting that, since radars for healthcare applications often work in indoor scenarios, the atmospheric loss can be neglected.

When the EM wave hits the target, it is reflected in many directions. The target behaves as if it was a source of EM waves itself, because the incident EM signal induces time-varying currents [75].

As a consequence, the power reflected back and received by the radar is influenced by the target and again by the channel. The target contribution can be expressed as a gain (less than unity), G_{targ}:

$$G_{targ} = \frac{4\pi\sigma}{\lambda^2} \tag{2.3}$$

where

- σ is the target radar cross section.

The radar cross section (RCS) is an important factor, expressed in square meters (m^2) and it depends on many parameters such as the physical size, the shape and the material of the target. The RCS is used to characterize the target thus it does not consider the transmitter and receiver characteristics nor the distance between target and radar (the RCS value is normalized by the power density of the incident wave at the target so that it does not depend on the distance).

The expression for the power at the radar receiver is:

$$P_r = \frac{P_t G_{targ} A_{fs} G_r}{L_t} \tag{2.4}$$

where

- G_r is the receiving antenna gain.

Of course, both A_{fs} and L_t must be considered again due to the round-trip path of the signal and can be written as $L_s = L_t{}^2$, whereas G_r may be in general different from G_t.

However, by considering $G_r = G_t = G$ Equation (2.4) can be written as in (2.5).

$$P_r = \frac{P_t G^2 \lambda^2 \sigma}{(4\pi)^3 L_s R^4}. \tag{2.5}$$

A radar engineer cannot change either the transmit/receive channel, or the target characteristics. However, a radar engineer can exploit the RRE to select and design the best radar components to fulfil the desired application. For this purpose, often the RRE is written in the form of (2.6).

$$R_{max} = \frac{P_t G^2 \lambda^2 \sigma}{(4\pi)^3 L_s P_{MDS}} \tag{2.6}$$

where

- R_{max} is the maximum range achievable by the radar
- P_{MDS} is the minimum detectable signal, i.e. by considering a certain suitable SNR, $P_{MDS} = P_t + SNR$.

Equation (2.6) is exploited to calculate the maximum range of the radar and of course it strongly depends on the SNR at the receiver that in turn depends on the characteristics of the receiver. In order to increase the R_{max}, it is usually preferable to increase the gain of the antennas or improve the sensitivity of the receiver in terms of a lower noise figure, rather than increase the transmitted power which is in a fourth root, e.g., to double the range, it is required to multiply the transmitted power by 16.

2.2 Detection Issues

As mentioned in the previous section, it should be noted that each received sample of radar data is a mix of the signal of interest and noise. If the sampled signal is not an echo, the sampled data includes only noise. The noise can be in turn a mix of thermal noise, receiver noise, and might also include air or ground clutter echoes, electromagnetic interference from other non-intentional EM sources, and hostile intentional jamming.

The term clutter refers to the echo from objects which are of no interest to the radar applications. As an example, if the radar mission involves the detection of moving ground vehicles, each echo from stationary objects such as mountains and vegetation or moving objects such as precipitation, is of no interest and it is considered an interfering contribution. On the other hand, for Earth mapping applications, sensing mountains and vegetation can be of interest as sensing precipitation is of interest for weather radars.

Instead, the term jamming refers to intentional interference due for example to electronic warfare. A radar system might be also subject to high-power EM waves coupled from its transmitter that would prevent the detection of targets and could severely damage the receiver's sensitive components; this is considered self-jamming.

Clutter can be distinguished from noise because noise amplitude does not depend on the transmitted signal level. In contrast, the amplitude of clutter reflection is proportional to the transmitted signal level. The noise bandwidth is wide, limited only by the receiver bandwidth, whereas the clutter is usually a narrowband signal. Noise and clutter can be separated also because noise is random and characterized by a constant average value, which does not depend on the range, while clutter is correlated between each different measurement [76].

The noisy received signals can be treated as additive random processes; thus, even if the target echo amplitude is entirely deterministic, the combined target-plus-interference signal is a random process [75]. This poses several challenges for target detection, which are categorized as detection issues.

We can consider two possible scenarios, the first whereby the data includes only interferences, the second whereby the data includes both interferences and returns from the targets of interest. The two scenarios are mutually exclusive, thus only one can be true. By referring to the first scenario as the null hypothesis H_0, and to the second one as H_1, it is possible to define the PDFs that a single sample y is not present $p_y(y|H_0)$ or is present $p_y(y|H_1)$.

The role of the detection logic in the radar processing unit is to examine each radar measurement and determine the best hypothesis, where the term "best" describes the hypothesis better accounting for the real scenario.

Should H_0 be the best hypothesis, then it is supposed that a target is not present, otherwise the target is present.

A very famous detection method is the Neyman–Pearson criterion. Based on a fixed probability of false alarm that can be tolerated by the system, P_{FA}, the purpose is to increase the probability of detection, P_D.

The thresholding detection procedure is illustrated in Figure 2.2. In detail, each data sample is compared to a threshold. If the sample overcomes the threshold, it is supposed to be due to the presence of the target plus the interference, as sample 50 in Figure 2.2; otherwise, it is supposed to be due only to the interference.

The choice of words "it is supposed to be" is not casual, because the assumption can be wrong. Indeed, should a peak due to a strong inference cross the threshold, this will lead to a false alarm. This is the case of sample 20 in Figure 2.2.

Figure 2.2: Threshold detection example.

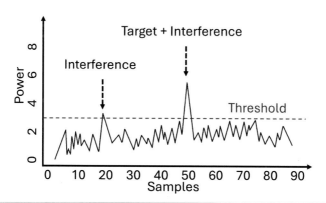

As a consequence, increasing the value of the threshold in turn decreases the probability of false alarm. On the other hand, a sample coming from a weak echo is not able to overcome the threshold; in this case the target is not detected thus causing a miss with probability $1 - P_D$.

Since for a defined radar system, increasing P_D also implies increasing P_{FA}, a trade-off should be selected according to the specific application and to the average number of false alarms per unit time tolerated by the system.

If the PDF of the scenario is available, the probability of false alarm can be calculated by considering the area under $p_y(y|H_0)$ from the threshold to infinite, whereas the probability of detection can be calculated by considering the area under $p_y(y|H_1)$ from the threshold to infinite. This is described in (2.7) and (2.8).

$$P_{FA} = \int_T^\infty p_y(y \mid H_0)\, dz \tag{2.7}$$

$$P_D = \int_T^\infty p_y(y \mid H_1)\, dz \tag{2.8}$$

where T is the threshold.

Figure 2.3 graphically shows this concept.

Figure 2.3: Gaussian PDF of the voltage y for the case of noise (left curve) and target plus noise (right curve) [75].

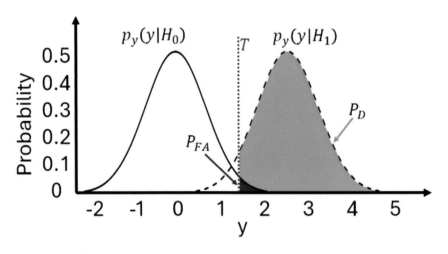

If the noise is a zero-mean Gaussian random process with variance s_n^2, whereas the signal of interest simply coincides with the mean value m, the target-plus-noise random process is also Gaussian with mean m and variance s_n^2.

The role of the radar engineer is to make the radar meet the performance specifications by increasing the P_D while decreasing P_{FA}. By observing Figure 2.3, shifting the two PDFs apart looks beneficial to this purpose. Indeed, the area under $p_y(y \mid H_0)$ decreases whereas the area under $p_y(y \mid H_1)$ increases. Since $p_y(y \mid H_0)$ cannot be moved, the only solution is moving $p_y(y \mid H_1)$ to the right that in turn means increasing the mean m, thus the SNR. Another possibility to improve the performance is making the two curves narrower, i.e. reducing their variance thus the noise power, s_n^2. Equivalently, reducing the noise power involves increasing the SNR, which is a fundamental result. Since the SNR is a key quantity affecting both P_D and P_{FA}, often the radar receiver characteristics are described by the ROC curve, which plots two of these three quantities, i.e. P_D, P_{FA}, and SNR with the third as a parameter and thus defines the expected performance of the receiver depending on the selected trade-off.

Often, the PDF is unknown or changes with time and space, thus the threshold selection becomes very challenging. A possible solution exploits dynamic thresholds based on different principles reported in the literature [75], [77]–[81]. A very famous example is the cell-average constant false alarm rate detector, which is based on tracking the main level of the interference to dynamically set the threshold with a fixed probability of false alarm.

2.3 Doppler Radars

One classic goal of radar detection is measuring the speed of a target. However, in recent years the growing interest in healthcare contactless monitoring has pushed the research effort towards Doppler radars, traditionally used only to detect speed, for biomedical applications as vital signs and movement sensors [82]–[84]. However, some basic principles are the same and are based on the Doppler effect. If there is a range variation between the radar and the target, in other words if there is a relative speed between them, and by considering f_c as the frequency of the transmitted signal, the frequency f_r of the received echo will be different from the transmitted one. This is a consequence of the Doppler effect.

By considering a monostatic radar, i.e. the transmitter and the receiver can be considered stationary and placed at the same location, if the target is moving with a radial speed component v in the direction of the radar, the theory of the special relativity allows to calculate the received frequency as in (2.9) [85], [86].

$$f_r = f_c \left(\frac{1 + v/c}{1 - v/c} \right) \tag{2.9}$$

where c is the speed of light.

From (2.9), it is worth noting that if the target is approaching the radar, the received frequency increases; otherwise, if the target is moving away from the radar, the received frequency decreases. A parallel can be drawn for the case of sound waves, when the sound from moving ambulance sirens or train whistles appears distorted for a subject stationary or moving at different speed.

The difference f_d between the transmitted and received frequencies is called the Doppler frequency or Doppler shift and can be calculated as

$$f_d = f_c \frac{2v}{c} \cos\vartheta = \frac{2v}{\lambda} \cos\vartheta \qquad (2.10)$$

where ϑ is the angle between the velocity vector of the target and the radar LOS.

Equation (2.10) shows that the Doppler shift is proportional to the relative velocity along the line of sight between radar and target, called the radial velocity $v\cos\vartheta$.

In addition to the basic function of speed detection, Doppler shifts can be exploited in radar systems to detect echoes from moving targets in the presence of much stronger reflections from stationary clutter. In this case, detecting the speed is not the goal but a means to separate moving from stationary targets. On the other hand, measuring the speed can be also useful to correct the range or angular measurement [87]. Very famous types of radars are continuous-wave radars, so called due to the time-continuous nature of the transmitted signal. Since they are usually exploited to detect Doppler shifts, they are known as Doppler radars. In the basic operation of a Doppler radar, a transmitter sends out a signal $s_{tx}(t)$:

$$s_{tx}(t) = cos\left(2\pi f_c t + \phi_0\right) \qquad (2.11)$$

where ϕ_0 is the residual phase.

In (2.11) the signal amplitude has intentionally not been considered. Indeed, although the signal amplitude is a key parameter to allow the detection, i.e. a minimum SNR should be ensured at the receiver, the Doppler detection is not based on amplitude information.

Part of the transmitted signal will be reflected by a target and detected by the radar receiver. A frequency shift relative to the transmitted signal due to the Doppler effect will be noticed in the received signal if the target has a non-zero radial velocity component v [5]. The received signal can be represented as:

$$s_{rx}(t) = cos\left(2\pi f_c t \pm 2\pi f_d t + \Delta\phi\right) = cos\left(2\pi f_c t \pm \frac{4\pi v t}{\lambda} + \Delta\phi\right) \qquad (2.12)$$

where $\Delta\phi$ appears because of the nominal detection distance (this concept will be clearer later) and accumulated phase noise. The plus sign corresponds to an approaching target whereas the minus sign corresponds to a receding target.

The term $\pm\frac{4\pi\nu t}{\lambda}$ is of interest, because it allows extraction of the Doppler frequency.

The scientific literature reports on different architectures to measure the Doppler frequency in continuous wave radars [4], [5], [9], [88], [89]. Among the different available topologies, the present text describes the use of a homodyne receiver with quadrature mixer to directly convert the received signals to the baseband. A simplified block diagram of a direct conversion receiver is shown in Figure 2.4.

Figure 2.4: Simplified block diagram of a homodyne receiver with quadrature mixer.

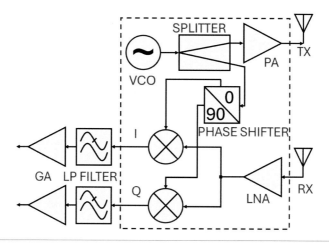

The signal is generated by a VCO or signal synthesizer and partly directed towards the receiving chain. Most of the signal is amplified and sent to the environment through an antenna and the reflected signal, if any, is again received by an antenna and amplified.

This architecture splits the signal in the receiving chain into two channels. The signal in the upper channel is mixed with the reference oscillator $cos\,(2\pi f_c t)$, whereas the signal in the lower channel uses a reference oscillator $sin\,(2\pi f_c t)$. The lower oscillator is therefore 90° out of phase with the upper one; this condition is referred to as being in quadrature with respect to the other one. The cascade of the IQ mixer and low-pass filtering section allows obtaining a signal whose frequency is proportional to the frequency difference between the transmitted and the received signals, thus highlighting the Doppler-dependent term. The in-phase and quadrature signals are reported in (2.13) and (2.14), respectively.

$$s_I(t) = \cos(\pm 2\pi f_d t + \Delta\phi) \tag{2.13}$$
$$s_Q(t) = \sin(\pm 2\pi f_d t + \Delta\phi). \tag{2.14}$$

From (2.13) and (2.14), it is straightforward to measure not only the target speed but also the direction because the frequency and the sign are known. This is required because positive velocities mean an approaching target and negative velocities mean a leaving target. A new, complex-valued signal $s_b(t)$ can be obtained from the in-phase and quadrature outputs, as shown in (2.15). This complex output allows independent measurement of amplitude and phase, thus it is possible to understand if the target is approaching/moving away from the radar, according to the velocity sign.

$$s_b(t) = s_I(t) + j s_Q(t) = e^{j(\Delta\phi \pm 2\pi f_d t)} = e^{j\left(\Delta\phi \pm 2\pi \frac{2\nu}{\lambda} t\right)}. \tag{2.15}$$

This can be visually represented by using an IQ diagram, as in Figure 2.5, where the angular speed is proportional to the Doppler frequency.

Figure 2.5: IQ diagram representing the case of (left) an approaching, (right) a receding target.

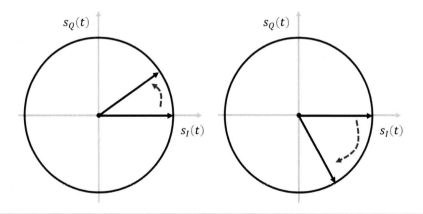

However, it is also possible to analyze the same received signal by following a different process. Indeed, (2.12) can be re-written as in (2.16):

$$s_{rx}(t) = s_{tx}\left(t - \frac{2d(t)}{c}\right) = \cos\left(2\pi f_c\left(t - \frac{2d(t)}{c}\right) + \phi_0\right) \tag{2.16}$$

where $d(t)$ is the distance of the target, where the dependance on the time t is due to the target movement.

$d(t)$ can be written as a nominal starting and fixed distance R_0 plus a time-varying distance shift $x(t)$. $s_{rx}(t)$ can be re-written as:

$$s_{rx}(t) = cos\left(2\pi f_c t + \frac{4\pi R_0}{\lambda} \pm \frac{4\pi x(t)}{\lambda} + \phi_0\right). \tag{2.17}$$

Therefore, the IQ signals are:

$$s_I(t) = cos\left(\frac{4\pi R_0}{\lambda} \pm \frac{4\pi x(t)}{\lambda} + \phi_0\right) = cos\left(\Delta\phi \pm \frac{4\pi x(t)}{\lambda}\right) \tag{2.18}$$

$$s_Q(t) = sin\left(\frac{4\pi R_0}{\lambda} \pm \frac{4\pi x(t)}{\lambda} + \phi_0\right) = sin\left(\Delta\phi \pm \frac{4\pi x(t)}{\lambda}\right) \tag{2.19}$$

where:

- $\frac{4\pi R_0}{\lambda}$ is constant because the initial range is constant
- ϕ_0 is constant because it represents the residual phase.

Therefore, both can be included in the constant term $\Delta\phi$, already seen in (2.12), whose meaning should be now clear.

In order to extract the phase $\varphi(t)$, the quadrature signals $s_I(t)$ and $s_Q(t)$ are combined by the arctangent operator.

$$\varphi(t) = tan^{-1}\frac{s_Q(t)}{s_I(t)} = \Delta\phi \pm \frac{4\pi x(t)}{\lambda}. \tag{2.20}$$

Since $\Delta\phi$ is nearly constant and useless for the instantaneous phase extraction, it is usually removed by subtracting the starting phase value.

The term $\varphi(t)$ is really useful because it is proportional to the target range shift, which can be now extracted and exploited to measure very small displacements such as the vital signs. Indeed, being related to a phase shift, this is a highly sensitive measurement.

As a matter of fact, the Doppler speed measurement is related to a phase change over time. By using a quadrature receiver, it is also possible to gather the direction of movement from the phase shift. The speed can be calculated from the phase by applying a Fourier Transform on the signal $s_b(t)$:

$$s_b(f) \approx e^{j\Delta\phi}\delta_D\left(f - \frac{2\nu}{\lambda}\right). \tag{2.21}$$

The function δ_D will be centered in $\frac{2\nu}{\lambda}$, i.e. the frequency spectrum shows a peak centered on the Doppler frequency.

However, it is worth noting that the phase information does not allow measuring the range of the target but only range shifts from the initial position.

Indeed, since the initial range of the target is R_0, the total number of wavelengths λ in the two-way path from radar to target and return is $\frac{2R_0}{\lambda}$. Each wavelength corresponds to a phase change of 2π radians, therefore, the total phase change in the two-way propagation path is the already known expression $2\pi \times \frac{2\,R_0}{\lambda} = \frac{4\pi\,R_0}{\lambda}$. The phase at range 0 starts from 0, afterwards the wavelength becomes equal to 2π and reset to 0 n times during the roundtrip path. The residual phase after $\frac{4\pi\,R_0}{\lambda}$ is $\Delta\phi$, which, as already mentioned, is commonly not useful and is being discarded.

Finally, the longer the observation time, the better the speed resolution, which is related to the frequency resolution of standard spectrum analysis method based on fast Fourier transform (FFT).

Whereas the ambiguity due to the phase discontinuity, i.e. the phase is limited within the interval $(-\pi, \pi)$ while the target movement is not, is solved by means of a phase un-wrapping processes, it is possible to unambiguously measure the velocity only if the phase change between two measurements is less than π. Therefore, the maximum measurable displacement $x(t)_{max}$ can be computed as:

$$\frac{4\pi x(t)}{\lambda} < \pi \Rightarrow x(t)_{max} = \frac{\lambda}{4} \tag{2.22}$$

and the maximum measurable velocity v_{max} can be expressed as:

$$\frac{4\pi\nu(t)\,T_m}{\lambda} < \pi \Rightarrow v_{max} = \frac{\lambda}{4T_m} \tag{2.23}$$

where T_m is the time interval between two measurements, thus it usually corresponds to the sampling time.

Let us note that the additional 2 at the denominator of (2.23) is a consequence of the round-trip path.

It is worth noting that the arctangent signal demodulation based on (2.20) can eliminate the optimum/null detection point problem by combining the I and Q signals in the baseband [90].

As mentioned in Section 1.2, the phase analysis in terms of displacement and Doppler detection considers the target as a single point or a parallel plane. However, in many cases, the target should be considered as a rigid or non-rigid body, with each sub-component moving independently, thus requiring a different kind of analysis. Whereas a rigid body has finite size, and each subcomponent moves without deformation, i.e. the distance between any two points of the body itself is constant during the movement, a non-rigid body is deformable thus changing its shape. As a matter of fact, treating the target as a rigid body is an idealization but makes the analysis easier compared to the case of non-rigid bodies [24].

These two cases are often referred to as micromovement and macromovement [91].

The term micro-Doppler signature refers to a distinctive expression of the target movement in terms of its Doppler components and it is represented in the joint time/Doppler domain. Since it is related to a specific characteristic of the target, it can be exploited for object identification, movement analysis, etc.

The term micro in micromotion refers to a large class of cases such as target oscillations, vibrations, rotations, swinging and flapping of small extents.

As it will be discussed in the last section, human body movement analysis is an important topic example benefiting from micro-Doppler research. Of course, it is also a complex field due to complex human motion due to articulation and deformability of the body.

The physical effects implied in the micro-Doppler were not discovered within radar research activities, but they were observed in coherent LADAR (laser detection and ranging) thus transmitting EM waves at higher frequencies in the optical spectrum. The feature of being coherent is important to preserve the phase information with respect to the EM signal generated by the LO.

In [24], a relationship is reported to calculate the maximum Doppler frequency shift if the target is vibrating with a vibration frequency f_v.

$$max\{f_D\} = \frac{2}{\lambda}f_v D_v \tag{2.24}$$

where D_v is the vibration amplitude.

It is worth noting that (2.24) implies that even in systems characterized by both low vibration rate and amplitude, employing microwave or millimeter wave radars, thus with very small wavelength, is very beneficial to detect the Doppler shifts induced by motion. Since the micro-Doppler effect is sensitive to the product $1/\lambda$, a radar engineer could operate at higher frequencies to observe tiny movements of the target, e.g., vibrations.

The radar architecture required to measure the time-varying frequency shift is the same employed for Doppler radars, but the required signal processing is different. Indeed, even though the Fourier transform is a great tool to analyze the frequency behavior of the target, it is not suitable for measuring time-varying frequency features.

As a matter of fact, the simultaneous analysis of time and instantaneous frequency domains is required. The term instantaneous frequency is often derived as a result of the time-derivative operation, which is suitable for monocomponent signals, i.e. a narrowband signal at any time. However, the target analysis as a rigid or non-rigid body involves multicomponent signals, i.e. signals characterized by multiple frequency sub-bands need to be combined over time to localize the energy distribution in both the time and frequency domains.

A suitable representation of both monocomponent and multicomponent signals is the spectrogram, i.e. a time vs. frequency representation of the signal. It can be calculated by using the squared magnitude of the short-time Fourier transform. The STFT performs several Fourier transforms in a short time and overlapped time window. This differs from the Fourier transform analysis which is applied on the entire signal.

Since the time window is limited, there is a tradeoff between time and frequency resolution. The larger the window size, the better the frequency resolution but the poorer the time resolution.

In addition to the time window size and type, another key parameter to obtain these Doppler-time patterns is the overlap factor [92].

As an example of micro-Doppler measurement, Figure 2.6 shows the Doppler time patterns related to a subject who is walking in front of the radar. From the first section of the figure it is possible to understand that the subject is approaching the radar due to the positive average Doppler components, while, after a turn around, the subject is moving away from radar due to the negative average Doppler components. In addition, the micro-Doppler signature highlights the micromovement of the target thus, for example from the first section in the figure, the lightest strips relate to the echo from the part of the body with the higher RCS, the torso, whereas the long alternating strips can be attributed to the swinging arms during the walking activity.

2.4 FMCW Radars

In the previous section, how a CW radar can be exploited to measure the speed and the target displacement was described, highlighting that it is

Figure 2.6: Example of a micro-Doppler signature of a walking subject.

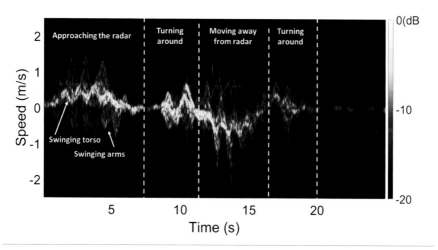

not possible to detect the target range. The next goal is to investigate how to detect both the range and speed of targets; to this aim, the key information behind the working mode, which will be described in the present section, is that the range measurement is being related to a frequency measurement.

A frequency-modulated continuous wave radar transmits a series of signals called chirps. A chirp is a sinusoidal signal whose frequency varies with time. It is possible to employ different modulation waveforms and, depending on the type of modulation, performance changes [93]. One of the most frequently used kinds of modulating signal is the sawtooth waveform. In Figure 2.7a, a signal linearly modulated by a sawtooth waveform is shown. A frequency vs. time plot is a convenient way to represent the time evolution of a chirp, as in Figure 2.7b. In detail, the modulating bandwidth is B, whereas the time duration of the chirp is T_c. The slope S of the chirp can be defined as the rate at which the chirp ramps up.

$$S = \frac{B}{T_c}.$$ (2.25)

The transmitted FMCW chirp, $s_{tx}(t)$ can be expressed as

$$s_{tx}(t) = cos\left(2\pi f_c t + \frac{\pi B}{T_c}t^2 + \phi_0\right).$$ (2.26)

Figure 2.7: (a) Amplitude vs. time and (b) frequency vs. time representation of a chirp.

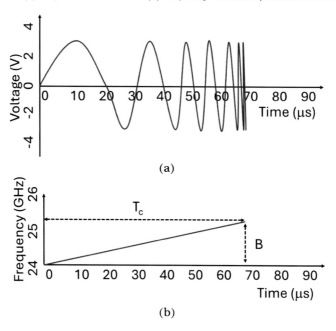

(a)

(b)

As for (2.11), the signal amplitude has been intentionally considered unitary to simplify the text. The instantaneous frequency can be calculated from the derivative of the argument $(f_T(t) = f_c + \frac{B}{T_c}t)$, whereas ϕ_0 will be neglected as for the case of Doppler radar.

If a target is present in the scenario, it can reflect the transmitted signal back to the radar, thus the received signal $s_{rx}(t)$ is a scaled and delayed version of the transmitted one.

$$s_{rx}(t) = s_{tx}(t - t_d) \approx \cos\left(2\pi f_c(t - t_d) + \frac{\pi B}{T_c}(t - t_d)^2\right) \qquad (2.27)$$

where $t_d = \frac{2R}{c}$ is the round-trip delay and it is usually a small fraction of T_c; indeed, for a target at 300 m and $T_c = 100\ \mu s$, $\frac{t_d}{T_c} = 2\%$. Again, the path loss has been neglected by maintaining the amplitude equal to 1.

The transmitted/received signal pair is graphically shown in Figure 2.8, where it is interesting to observe the beat frequency f_b obtained as frequency difference between the frequency of the transmitted signal by that of received one.

Figure 2.8: Transmitted (red curve), received (black curve) and difference between transmitted and received frequency (blue curve).

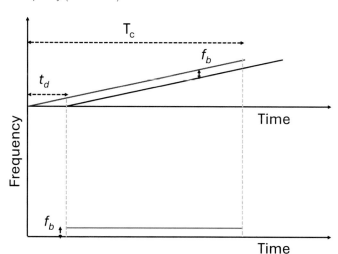

If the reader thought to move the received signal right, f_b would increase whereas if the reader thought to move the received signal left, f_b would decrease. Moving the received signal right is equivalent to having a farther target, whereas moving the received signal left is equivalent to having a nearer target. In other words, for an FMCW radar, measuring the range is equivalent to measuring the frequency difference f_b, because f_b is directly proportional to the target range according to (2.28).

$$f_b = \frac{2\,S\,R}{c} = 2R\frac{B}{c\,T_c} \Rightarrow R = \frac{c\,T_c}{B}\frac{f_b}{2}. \tag{2.28}$$

As for the CW/Doppler radar case, by performing a frequency difference, e.g., by employing a homodyne receiver with quadrature mixer, information about the target can be obtained. However, for an FMCW radar, a single stationary object in front of the radar will produce a low frequency signal or beat signal which is ideally a constant frequency tone. The basic architecture of an FMCW radar is the same as Figure 2.3, with some adjustments for the cutoff frequency of the filters and by considering that the VCO input is not a constant voltage, but it is a sawtooth signal.

Multiple objects in front of the radar involve multiple reflected chirps at the RX antenna. A frequency spectrum of the beat signal, typically performed by means of an FFT, will reveal multiple tones, the frequency of each being proportional to the range of each object from the radar. A graphical view of this case is shown in Figure 2.9 where the frequency vs. time and the spectrum of the beat signal are represented.

Figure 2.9 recalls that multiple targets can be present in the scenario and in this case their range separation will be required. The range resolution is the radar ability to separate two targets in the same scenario but at different ranges. Analyzing the range resolution is very important to understand what the

Figure 2.9: (a) Transmitted (red curve), received (black curve) and difference between transmitted and received frequency (blue curve) for the case of multiple targets and (b) frequency spectrum of the related beat signal.

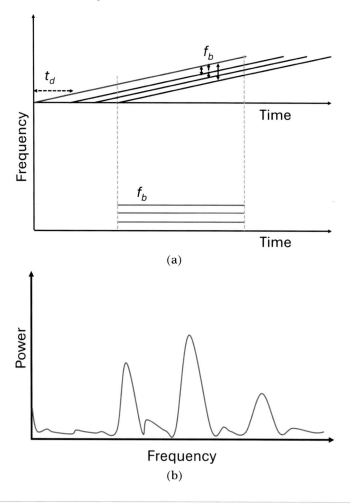

minimum distance between two targets is to allow their separation, i.e. the two targets are not detected as a single target. The resolution δR can be calculated from the derivative δf_b of f_b as in (2.29).

$$\delta f_b = 2\, \delta R \frac{B}{c\, T_c}. \tag{2.29}$$

Since $\delta f_b \approx \frac{1}{T_c}$, (2.29) can be written as

$$\delta R = \frac{c}{2\, B} \tag{2.30}$$

which is the expression of the range resolution.

From (2.30), it is evident that the higher the bandwidth, the better the resolution, thus increasing the bandwidth is very beneficial for radar engineers.

On the other hand, the range accuracy is the radar capability to measure the correct range of the target. According to [94], the radar accuracy σ_R can be calculated by exploiting (2.31).

$$\sigma_R = \frac{c}{3.6\, B\sqrt{2\, SNR}}. \tag{2.31}$$

From (2.31), it is worth noting that the accuracy is a fraction of the range resolution and, for a fixed range resolution, improving the SNR is beneficial to improving the accuracy. Figure 2.10 aims to clarify these concepts by reporting (a) an example related to the range accuracy whereby there is only one target in the scenario and the purpose is to find the correct value and (b) an example related to the range resolution whereby two targets are in the scenario, but they are detected as a single target due to the poor range resolution.

Section 2.1 explained how the radar range equation can be exploited to find the maximum detectable range. However, for FMCW short range radars, sometimes the limit imposed by the radar range equation is not reached and the maximum range is limited by radar parameters such as the modulation bandwidth, chirp duration and the sampling frequency set during the ADC conversion of the beat signal. This is a direct consequence of the digital analysis of the radar returns that are typically acquired by exploiting an ADC. Since the range detection relies on a frequency measurement, the maximum detectable range R_{max} in turn depends on the maximum detectable frequency $f_{b\,max}$, which is limited by the sampling frequency F_s in a digital system, i.e. $f_{b\,max} < \frac{F_s}{2} \cdot f_{b\,max}$ can be in turn calculated by re-elaborating (2.28) as follows.

$$f_{b\,max} = 2 R_{max} \frac{B}{c\, T_c}. \tag{2.32}$$

Figure 2.10: (a) Range accuracy: only one target in the scenario with the task of finding the correct range and (b) range resolution: two targets in the scenario (black and red solid curves) are detected as a single target (blue dashed curve) if the resolution is not appropriate.

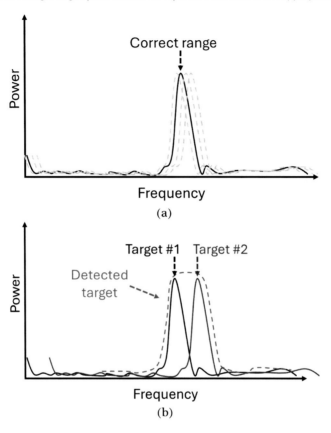

Therefore

$$R_{max} = \frac{F_s\, c\, T_c}{4B}. \tag{2.33}$$

So far, the FMCW radar working principle has been described by analyzing only a single chirp. However, typical FMCW configuration consists of a sequence of chirps followed by idle time. Chirp sequences are often packaged in groups, called frames.

The typical pulsed radar terminology includes the term pulse repetition time, PRT and pulse repetition frequency, PRF. These terms are commonly used for FMCW radars to indicate the inter-chirp period i.e. the time between the start of two consecutive chirps and the chirp rate, respectively.

In general, the PRT and T_c do not coincide. Indeed, between consecutive chirps, an idle time can be present thus increasing the duration of the PRT compared to T_c. This is very important for the remainder of the text because the speed detection is based on the effective temporal distance between the consecutive chirps. However, to simplify the text, the idle time is considered to be equal to zero and thus the PRT and T_c coincide.

A graphical representation of the chirp sequence is shown in Figure 2.11.

Figure 2.11: Graphic representation of the chirp sequence.

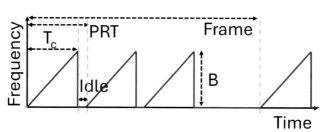

An FFT is performed on every chirp and the result is subsequently stored in a row/column of a matrix. The time domain within a chirp is usually referred to as fast time. Sequences of chirps fill a data matrix and the time domain across multiple chirps is referred to as slow time. The data matrix can be graphically represented to provide the range vs. time and assumes the name range profile.

Let us consider M chirps, each composed by N bins. Therefore, M is the number of chirps-per-frame and N is the number of samples-per-chirp. This is graphically represented in Figure 2.12 where (a) the spectrum of multiple downconverted chirps (one spectrum for each chirp) and (b) the corresponding range matrix are represented.

In Figure 2.13, an example of a range profile from measured data is reported, representing a person walking from the starting range of 3 m to the final range of 1m.

As for the case of Doppler radars, a quadrature receiver can be exploited to demodulate the received signal thus obtaining both the module and the phase of the echo signal, $S_b(t) = S_I + j\, S_Q$. Again, this can be beneficial for detecting the target speed and measuring small displacements, e.g., the vital signs. The

Figure 2.12: (a) Spectrum of multiple downconverted chirps (one spectrum for each chirp) and (b) the corresponding range matrix are represented.

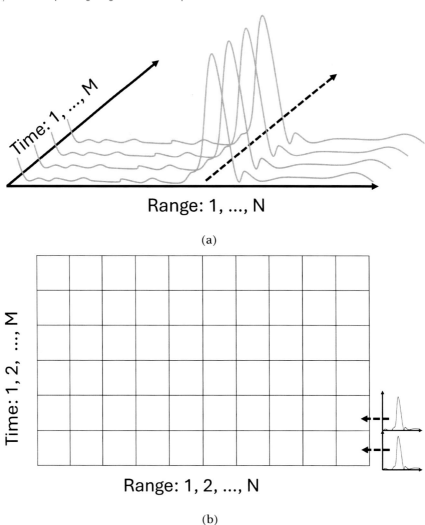

Range: 1, ..., N

(a)

Range: 1, 2, ..., N

(b)

in-phase and quadrature signals are reported in (2.34) and (2.35).

$$S_I(t) = \cos\left(\frac{2\pi \, B \, t \, t_d}{T_c} + 2\pi \, f_c \, t_d\right) \tag{2.34}$$

$$S_Q(t) = \sin\left(\frac{2\pi \, B \, t \, t_d}{T_c} + 2\pi \, f_c \, t_d\right). \tag{2.35}$$

Figure 2.13: Range profile from measured data representing a person walking from the starting range of 3 m to the final range of 1 m. (a) Top view and (b) perspective view.

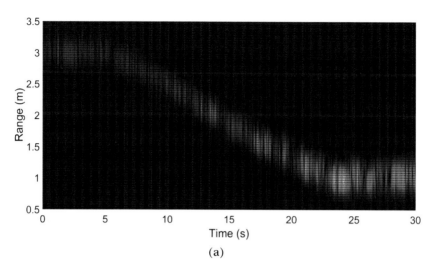

(a)

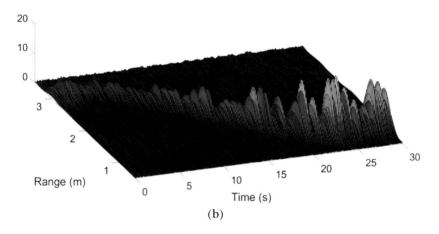

(b)

Or alternatively

$$S_b(t) = e^{j\left(\frac{2\pi\, B\, t\, t_d}{T_c} + 2\pi\, f_c\, t_d\right)}. \tag{2.36}$$

Equation (2.36) is re-written in (2.37) as a function of fast and slow time, t and τ, respectively. τ can be considered as the center time of each chirp. As for the case of Doppler radars, the range can be written as the sum of a nominal starting detection distance R_0 and the time-varying distance shift $x(\tau) = v\,\tau$.

$$S_b(t,\tau) = e^{j\left(\frac{4\pi B t (R_0 + v\tau)}{cT_c} + \frac{4\pi f_c (R_0 + v\tau)}{c}\right)}. \tag{2.37}$$

After some re-arrangement:

$$S_b(t,\tau) = e^{j\left(\frac{4\pi B R_0}{c\,T_c}t\right)} e^{j\left(\frac{4\pi B v}{cT_c}t\tau\right)} e^{j\left(\frac{4\pi f_c R_0}{c}\right)} e^{j\left(\frac{4\pi f_c v}{c}\tau\right)}. \tag{2.38}$$

Since data are managed digitally, it is useful going from the analog to digital domain, thus $t = n/f_s$, where n is the n_{th} sample in a chirp and f_s is the sampling rate [93]. For the mth chirp (m chirps-per-frame) it is possible to write $\tau_m = mPRT + t = mPRT + n/f_s$, thus from $S_b(t,\tau)$ it is possible to obtain $S_b(k, m)$.

$$S_b(n,m) = e^{j\left(\frac{4\pi B v n\, m\, PRT}{c\,T_c f_s}\right)} e^{j\left(\frac{4\pi B v n^2}{cT_c f_s{}^2}\right)} e^{j\left(\frac{4\pi f_c v m\, PRT}{c}\right)} e^{jn\left(\frac{4\pi B R_0}{cT_c f_s} + \frac{4\pi f_c v}{cf_s}\right)} e^{j\left(\frac{4\pi f_c R_0}{c}\right)}$$

| Range Doppler coupling | Higher order term | Inter-chirp Doppler phase | Range + intra-chirp Doppler phase | Constant |

$$\tag{2.39}$$

The higher order term can be ignored because usually in FMCW radar systems $\frac{4\pi B v n^2}{cT_c f_s{}^2} = 1$.

There are two time-dependent parameters: n and m.

As already mentioned, a frequency domain analysis can be performed to highlight the frequency components corresponding to the target range. Therefore, an FFT is usually applied to n first (one FFT for each chirp). In order to move from the time to frequency domain, n becomes k. The frequency is converted to range, thus k represents a range bin. The frequency domain expression $S_b(k, m)$ of $S_b(n, m)$ is reported in (2.40).

$$S_b(k,m) \approx e^{j\left(\frac{4\pi f_c R_0}{c}\right)} e^{j\left(\frac{4\pi f_c v m PRT}{c}\right)} \delta\left(k - N\frac{2B/T_c R_0 + 2f_c v + 2B/T_c v m PRT}{c \cdot f_s}\right). \tag{2.40}$$

It is worth noting that the constant term and the inter-chirp Doppler phase do not depend on n. $S_b(k, m)$ represents the range profile for the mth chirp, whereas a δ is used to simplify the text [93]. The frequency peak of the range profile, k_{peak} is located at

$$k_{peak} = N\frac{2B/T_c R_0 + 2f_c v + 2B/T_c v m PRT}{cf_s}. \tag{2.41}$$

If the target speed is equal to zero, it can be simplified to the already known (2.28):

$$\frac{k_{peak} f_s}{N} = f_b = \frac{2B/T_c R_0}{c}. \tag{2.42}$$

Therefore, the range can be calculated from k_{peak}. If $v \neq 0$ the target's speed should be assumed to be constant during the sequence of chirps to ignore the range/Doppler coupling. The calculated range might be corrected after calculating the speed (not required for slow targets).

The next step is obtaining the speed of the target. Equation (2.40) can be re-written as follows

$$S_b\left(k, m\right) = \beta e^{j\left(\frac{4\pi f_c R_0}{c}\right)} e^{j\left(\frac{4\pi \cdot vmPRT}{c}\right)} \tag{2.43}$$

where

$$\beta = \delta\left(k - N\frac{2B/T_c R_0 + 2f_c v + 2B/T_c vmPRT}{c \cdot f_s}\right).$$

Again, a frequency domain analysis can be performed to highlight the frequency peaks along m, in this case corresponding to the target speed. Therefore, a second FFT is applied to m (one FFT for each range bin), thus obtaining

$$S_b\left(k, \gamma\right) \approx \beta e^{j\left(\frac{4\pi f_c R_0}{c}\right)} \delta\left(\gamma - M\frac{2f_c vPRT}{c}\right) \tag{2.44}$$

where γ is a Doppler bin.

The frequency peak of the range profile, γ_P is located at

$$\gamma_P = M\frac{2f_c vPRT}{c} \tag{2.45}$$

and the Doppler speed of the target will be:

$$v = \frac{c}{2Mf_c PRT}\gamma. \tag{2.46}$$

A 2D FFT can be directly implemented to obtain range and Doppler information within milliseconds, which is an essential requirement for real-time applications that require high update rate and low latency.

After the second series of FFTs, it is possible to obtain the related 2D range-velocity graph, called a range-Doppler map. Let us note that one speed FFT is available for each range bin, i.e. the range-Doppler map is built by placing side by side all the speed FFTs in the matrix for each range bin. The speed term is the same seen for the Doppler radar, i.e. after the first FFT, and for each range bin the radar behaves as a Doppler radar. Indeed (2.45) is equivalent to (2.21). An example of a range-Doppler map is reported in Figure 2.14, where a target placed at a distance of 1 m is moving at a speed of 9 m/s.

On the other hand, as for the case of Doppler radars, the phase of the received signal can be exploited to extract the target displacement $x(\tau)$. Indeed (2.36) can be re-written as:

$$S_b\left(t, \tau\right) = e^{j\left(2\pi f_b t + \frac{4\pi f_c \left(R_0 + x(\tau)\right)}{c}\right)}. \tag{2.47}$$

Figure 2.14: Range-Doppler map showing a target placed at a distance of 1 m moving at a speed of 9 m/s.

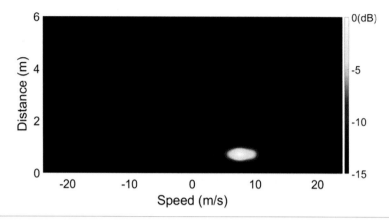

Again, it is worth noting that the term $\frac{4\pi \, f_c \, R_0}{c}$ is a constant and thus it can be replaced with $\Delta\phi$.

$$S_b\left(t, \tau\right) = e^{j\left(2\pi f_b t + \frac{4\pi f_c x(\tau)}{c} + \Delta\phi\right)}.$$ (2.48)

Therefore, the quadrature signals $s_I\left(\tau\right)$ and $s_Q\left(\tau\right)$ are combined by the arctangent operator and exploited to extract $x(\tau)$:

$$\varphi\left(t\right) = tan^{-1}\frac{s_Q\left(\tau\right)}{s_I\left(\tau\right)} = \Delta\phi \pm \frac{4\pi x(\tau)}{\lambda}.$$ (2.49)

Since $\Delta\phi$ is nearly constant and useless for the instantaneous phase extraction, it is usually removed by subtracting the starting phase value.

Therefore, to recap, fast FMCW radars can detect the objects' range by measuring the frequency shift of a single received chirp. If the target is moving, the beat frequency will be slightly modified but this tiny shift cannot be noticed from the frequency spectrum. By analyzing the phase of the received signal, the target range shift and equivalently the target speed can be obtained.

A frequently raised question is whether it is possible to exploit the frequency shift instead of the phase shift to detect the range shift?

A good way to answer this question is by means of a practical example. Let us consider a 60 GHz FMCW radar with 2 GHz bandwidth changing its target range by $\Delta x = 1.25$ mm. Let us calculate both the frequency and phase shifts. According to (2.28), the frequency shift Δf_b is $\Delta f_b = 2\Delta x \frac{B}{c\,T_c} = 16.7$ Hz. On the

Figure 2.15: Beat frequency vs. time spectrum of the beat frequency and related IQ graph for the case of a range shift from (a) 0 mm to (b) 1.25 mm.

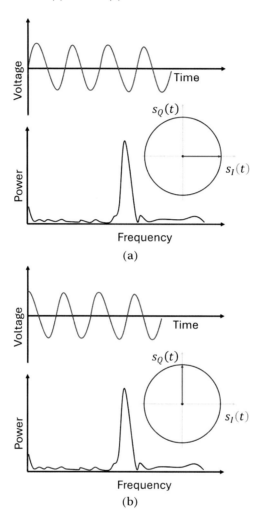

other hand, according to (2.49), the phase shift $\Delta\varphi$ is $\Delta\varphi = \frac{4\pi\Delta x}{\lambda} = 1.57$ rad $= \frac{\pi}{2} = 90°$.

Whereas 16.7 Hz is a very small frequency shift to be detected over an f_b in the order of tens of kHz or MHz, 90° is an enormous phase shift and can be easily detected. Therefore, the phase of the down converted signal is very sensitive, as for the case of Doppler radars, and can be exploited to detect very small displacements, like those related to the vital activity. A graphical example is shown in Figure 2.15, where the 1.25 mm range shift cannot be observed from the beat frequency spectrum, whereas it is clear from the phase rotation.

As for the case of the Doppler radar, it is possible to unambiguously measure the velocity only if the phase change between two measurements is less than π. Therefore, the maximum measurable displacement can be computed with (2.22) as for the case of the Doppler radar.

The maximum measurable velocity v_{max} differs from (2.23) because the time interval between two measurements T_m should be replaced by the inter-chirp time, the PRT or, assuming the idle time null, T_c.

$$\frac{4\pi\nu(t)\,T_c}{\lambda} < \pi \Rightarrow v_{max} = \frac{\lambda}{4T_c}. \tag{2.50}$$

It is worth noting that, according to (2.33), increasing the chirp duration increases the radar maximum range, but from (2.50) it decreases the maximum unambiguous speed [95]. Therefore, as in many engineering problems, a trade off must be managed.

For an FMCW radar, evaluating the speed resolution is very important. Indeed, if multiple targets with the same range but different speed are present in the scenario, they can be resolved only if the speed resolution δv is enough. As for the case of the Doppler radar, the speed resolution depends on the measurement time, thus in this case, it depends on the active duration of the frame.

$$\delta v = \frac{\lambda}{2\,MT_c}. \tag{2.51}$$

As for the case of the range accuracy, the speed accuracy is a fraction of the speed resolution and depends on the SNR.

$$\sigma_v = \frac{\lambda}{3.6\,MT_c\sqrt{SNR}}. \tag{2.52}$$

2.5 Multiple-input Multiple-output Radars

FMCW radars demonstrate premium performance in terms of high-resolution range detection. However, the target precise location greatly benefits from the additional capability to measure the angle-of-arrival of the received echo over the azimuthal and the elevation planes, which is the main feature of MIMO radars. The term MIMO stands for multiple-input multiple-output and affects both the radar architecture and operating mode.

The main additional characteristics of MIMO radars rely on peculiar hardware architectures that, as the acronym suggests, involve multiple transmitting and receiving sections, transmitting orthogonal waveforms.

Other options to measure the AoA involve mechanically scanning antenna, or electronically scanning antennas.

A typical architecture of a MIMO radar equipped with two transmitters and four receivers is shown in Figure 2.16.

Each transmitting channel requires frequency multiplication and amplification stages, and the corresponding antenna, whereas different receiving stages have the task to amplify and down-convert the received signal. A processing unit is usually in charge of generating the required transmitted waveform and processing the down-converted signals. The reason for employing multiple transmitters and receivers will become clearer.

Figure 2.16: Typical 2T × 4R MIMO radar architecture [44].

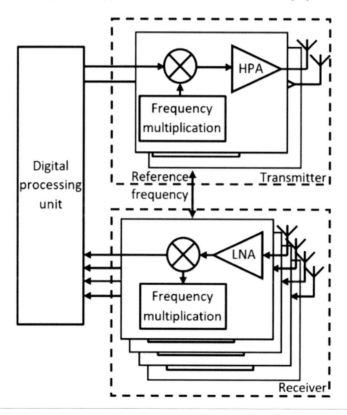

In addition to the angle detection capabilities, MIMO radars exhibit improved SNR compared to a SISO system due to the higher number of transmitter/receiver pairs.

The MIMO radar theoretical description implies some hypothesis to be true [44], [96]. In order to ensure constant propagation properties relative to the AoA of the received signal, the transmission medium might be isotropic and linear. This allows treating the received signals as the linear superposition of the signal wave fronts. The target might be in the far-field thus the received signals can be considered as parallel to each other. A rule of thumb to check the subsistence of the far-field condition can be found in [97].

$$R_{FF} > 2\frac{D^2}{\lambda}$$
(2.53)

where

- R_{FF} is the minimum range to ensure the far-field condition
- D is the maximum antenna dimension.

Moreover, the assumption of an AWGN channel ensures that the noise is not correlated, and the narrowband assumption guarantees that the characteristics of the received signals do not vary with the frequency.

Let us consider a target reflecting an EM wave $s\left(t\right)$ written as

$$s\left(t\right) = e^{-j[2\pi f_c t + \beta t]}$$
(2.54)

where β is the signal phase, while the amplitude has been set equal to 1 to simplify the text.

Let us consider now that $s\left(t\right)$ is directed towards an antenna array composed by m receiving antennas with $m = [1, 2, \ldots, M]$, separated by a distance d, as shown in Figure 2.17. As a matter of fact, the antenna array represents the MIMO radar receiving antennas.

The signal received by the first element $s_1\left(t\right)$ is a time shifted version of $s\left(t\right)$ and it can be written as

$$s_1\left(t\right) = e^{-j[2\pi f_c\left(t-\tau_d\right)+\beta\left(t-\tau_d\right)]}.$$
(2.55)

It is worth noting that the delay τ_d is proportional to the single-hop distance and not to the round-trip range R of the previous sections. This is only because in the current description, the signal starts from the target and not from the transmitter.

Figure 2.17: Uniform linear array (ULA) configuration [44].

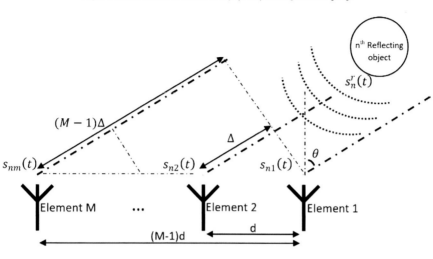

By considering a signal incidence angle θ and the reflecting object closer to the first antenna, the signal arrives at the receiving elements at different times, due to the additional distance $\Delta = d\sin\theta$ to be covered between consecutive elements.

By considering the first element as a reference, the signal at the m_{th} element, $s_m(t)$, undergoes an additional phase shift $e^{j(m-1)\frac{2\pi}{\lambda}d\sin\theta_n}$ with respect to $s_1(t)$. It depends on the distance d from the first element. $s_m(t)$ can be written as in (2.56).

$$s_m(t) = s_1(t)\, e^{j(m-1)\frac{2\pi}{\lambda}d\sin\theta} = e^{-j[2\pi f_c(t-\tau_d)+\beta(t-\tau_d)]}e^{j(m-1)\frac{2\pi}{\lambda}d\sin\theta} \, . \qquad (2.56)$$

By considering n reflecting objects with $n = [1, 2, \ldots, N]$ generating n reflected signals, (2.56) can be generalized as

$$s_{nm}(t) = s_n(t)\sum_{n=1}^{N} e^{-j(m-1)\frac{2\pi}{\lambda}d\cdot\sin\theta_n} \qquad (2.57)$$

$\begin{bmatrix} 1 & e^{j\frac{2\pi}{\lambda}d\cdot\sin\theta_n} & \ldots & e^{j(M-1)\frac{2\pi}{\lambda}d\cdot\sin\theta_n} \end{bmatrix}$ is the so-called steering vector.

As for the case of speed and range detection, for Doppler and FMCW radars, respectively, the phase is the key element for the AoA detection. Indeed, the steering vector is used to find the angular position of the target.

$\varphi_n = \frac{2\pi}{\lambda} d \sin \theta_n$ being the phase shift, the angular position of target n can be written as in (2.58).

$$\theta_n = \sin^{-1}\left(\frac{\varphi_n \lambda}{2\pi d}\right). \tag{2.58}$$

As for the case of Doppler and FMCW radars, an FFT can be applied to the different receiving elements to isolate the target angular position.

A good choice of d, in the case of uniform array, is $d = \lambda/2$, because since the phase is limited within the interval $[-\pi, \pi]$ it allows having the maximum FoV of $\pm\pi/2$ avoiding ambiguities in the angle estimation.

This key result can be exploited by using different antenna arrangements. Considered here is a uniform linear array (ULA) configuration, where M elements are equally spaced on a row. On the other hand, a uniform rectangular array (URA) is also possible [96]. In URAs the elements are arranged both horizontally and vertically. This two-dimensional array allows detection of both the azimuth and elevation of targets. Different options exist to arrange the elements. In Figure 2.18, three different two-dimensional centrosymmetric array configurations are shown.

The additional detection capabilities of URA, compared to ULA, involve a higher complexity of the steering matrix and consequently of the AoA processing.

From (2.57), it should be clear that the higher the number of antennas, the better the radar performance. Indeed, having a higher number of receiving elements increases the lengths of the steering vector and in turn increases the

Figure 2.18: Three different centrosymmetric URA configurations [44].

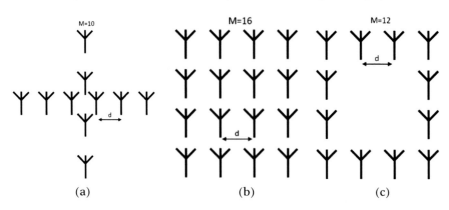

angular resolution. This can be better observed from (2.59) where the minimum angular distance $\Delta\theta_{min}$ required to separate two targets in the case of an equally spaced array is reported, i.e. the angular resolution.

$$\Delta\theta_{min} = \frac{\lambda}{M \, dcos\theta}. \qquad (2.59)$$

By considering M equally spaced antennas separated by a distance $d = \lambda/2$, and for a boresight view, i.e. $\theta = 0$, the angular resolution becomes $2/M$; thus, to double the angular resolution, the number of receivers should be doubled with a consequent adverse impact on the size and the complexity of the hardware architecture.

In order to obtain superior performance in terms of angular resolution without overloading the radar receiver, the number of virtual antennas can be increased.

The term array of virtual antennas refers to an array of antennas having the same performance of a physical array but having potentially a different number of physical elements.

The following example will show that the number of virtual antennas of a MIMO radar with M_{TX} transmitter and M_{RX} receivers is the same as a SIMO radar with one transmitter and $M_{TX} \times M_{RX}$ receivers. This feature provides a cost-effective way to improve the angular resolution.

Let us consider the 2×4 MIMO radar, where 2×4 is a compact version of $M_{TX} \times M_{RX}$, shown in Figure 2.19.

By considering the transmitter TX1 transmitting first, the phase shifts at the receiving elements referred to RX1 are $[0, \varphi, 2\varphi, 3\varphi]$. By considering the transmitter TX2, since it is $4d$ away from TX1, the phase shifts at the receiving elements referred to RX1 are $[0 + 4\varphi, \varphi + 4\varphi, 2\varphi + 4\varphi, 3\varphi + 4\varphi] = [4\varphi, 5\varphi, 6\varphi, 7\varphi]$, due to the additional path of length $D = 4d \, sin(\theta)$ compared to TX1.

By concatenating the two phase sequences from the two transmitters, the sequence $[0, \varphi, 2\varphi, 3\varphi, 4\varphi, 5\varphi, 6\varphi, 7\varphi]$ is obtained, which is the same sequence for the case of a SIMO radar with 8 receivers. An 8-element virtual array has been created, as shown in Figure 2.19.

However, the received signals should be separated depending on the transmitter to reconstruct the total steering vector, thus involving orthogonality between the transmitting channels. Several approaches have been investigated in the scientific literature as those based on time division multiplexing (TDM), frequency division multiplexing (FDM) or binary phase modulation (BPM) [93], [98].

Figure 2.19: 2×4 MIMO radar array [44].

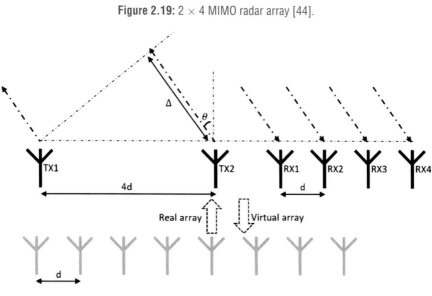

3

Applications of Radars for Healthcare

In this section, recent contributions concerning the applications of radar for healthcare are briefly described. Since many contributions have been reported in the scientific literature and great progress has been achieved, the contributions described in this section are mainly limited to the last 5 years to provide an up-to-date overview with the aim to stimulate the interest and the development of new activities in the field.

Great care should be applied to the ethical aspects involving the use of electromagnetic technologies on humans [99]. Every EM-based device must comply with the current national regulations in terms of emitted radiations, to ensure safe radiation levels both in domestic and medical environments and EM compatibility with other devices. Particular attention should be paid to the special requirements in the case of pathological conditions and training procedures for medical personnel.

As mentioned in Chapter 1, maintaining user comfort is one of the radar key strengths, together with not raising privacy concerns.

3.1 Vital Sign Detection

Vital sign detection represents one of the main purposes of radars for healthcare applications, and it is mainly based on the phase analysis described in the previous section. However, since contactless radars operating at microwave and mmWave frequencies are characterized by a short penetration depth, they can mainly detect skin surface movements. As a consequence, what the radar detects is the movement of a body induced by breathing and heart activity, which can occur in different parts of the body, such as the thorax or the abdomen.

The diaphragm extension goes from 1–2 cm during normal respiratory activity to 10 cm during deep activity [4]. First the oxygen O_2 passes through the upper respiratory system, the trachea and the bronchi and reaches the lungs; bronchi fork to alveoli inside the lungs, which manage the gas exchange with the blood; thereafter the carbon dioxide (CO_2) is finally exhaled [44].

By following a similar principle, the heart pumps the bloods through periodic contractions, thus inducing a motion of the skin surface. The circulatory system is in charge of absorbing O_2 and releasing CO_2, according to the inhalation/exhalation paths just described. This activity is scanned by the heart contractions that hit the chest wall, creating a measurable displacement. For the case of the heart activity, the heartbeat rate goes from 50 to 90 beats-per-minutes [44].

The scientific literature reports several examples of vital signs detection for both infants and adults, where the case of infants is often limited to sleep monitoring [100].

One of the main challenges occurring in real scenarios concerns the case of vital sign detection when the radar is not stationary. Indeed, by recalling that the radar measures relative movements, the tiny movement of the chest can be completely hidden by the radar movement itself. One practical occurrence takes place when the radar is mounted on a moving platform as a drone for rescue operations [101].

Decoupling the movement of interest from the radar motion itself often requires additional sensors to track the radar movements. In [66], this task has been accomplished without any additional sensor but by extracting the information concerning the radar movement from the echo coming from stationary targets which indeed is the mirror of the radar movement itself.

The other side of the coin is the case of vital sign detection when the target is moving, the so-called random body motion issue. As a matter of fact, in this case the target movement can also superimpose the tiny vital sign displacement. In [102], the heartbeat signal was detected during random body movement, based on joint blind source separation and a 4D imaging radar. It was validated by testing different types of possible real movements, showing that the heartbeat can be measured with a heart rate accuracy of 92.2%.

In many cases, the capability of radar to measure vital signs is useful to separate human from non-human targets. As an example, [13] proposes an automatic approach to identify humans in cluttered scenarios without the need for thresholding stages. The results are shown in a graphical representation denominated range-breathing graph, which quickly shows the human position for different ranges. The experimental analysis was performed at different frequencies, i.e. 24 GHz and 122 GHz.

Separating humans from non-humans can also be a useful requirement for longer ranges. In [103], the authors proposed a technique to increase the detection range by improving the radar sensitivity. In detail, they apply RF leakage/coupling cancellations in a dual-PLL low-IF Doppler radar. They experimentally demonstrated the vital signs detection feasibility of a subject 4 m away from radar by transmitting only −31 dBm of power.

As a matter of fact, the term vital signs includes both respiratory and heartbeat signals. However, extracting the heartbeat is very challenging due to the simultaneous presence of the respiratory signal which can have a similar frequency and higher SNR. Many contributions report on procedures for accurately extracting the heartbeat. In [104], the beat-to beat interval and the cardiac timing have been estimated from the single-band frequency envelogram by decoding the most likely sequence of states. Band pass filtering has been applied by means of a hidden semi-Markov model to measure the cardiac timing with high accuracy.

In [105], the authors designed a multi-band radar working at 80 GHz and 160 GHz, starting from a single 50 GHz LO. They implemented four channels, whereas the first two are used for a MIMO 80 GHz radar, the others work at 160 GHz and have an integrated wideband 6 dBi micromachined on-chip antennas to improve the range resolution. The system, working in FMCW mode demonstrated to detect the heartbeat rate of a subject with high Doppler resolution.

Heartbeat and breathing signals have been accurately separated also by exploiting differential operations applied to the measurement data together with autocorrelation techniques [106]. Employing the autocorrelation is a very common procedure to highlight periodic signal components. In [106] the authors also detected the heartbeat in the presence of close harmonic interferences and demonstrated the procedure effectiveness with an experimental analysis.

The heartbeat has been also separated from the breathing signal by exploiting the additional features in terms of angle separation of MIMO CW radars for multiple targets simultaneously [107]. To separate different subjects, a 2D digital beamforming techniques has been developed, first by finding the subjects' positions, then extracting the chest motion from the 2D image of the scenario, thus the vital signs. It is very interesting that the authors were able to focus the beam on the subjects' chests to improve the vital sign detection.

MIMO radars have also been exploited to isolate the respiratory signals of two close subjects, i.e. separated by less than 1 m [108]. The authors used an independent component analysis with the JADE algorithm together with the direction of arrival information and based the final detection on an SNR-based algorithm.

The concurrent angular separation and vital sign detection has also been demonstrated by exploiting an SISO radar [109]. In this case, transmitting and receiving antennas with frequency scanning capabilities were employed.

The task of detecting vital signs from different angles required to study the possible vital signs changes depending on the subject's orientations. To this aim, in [110], a distributed radar network has been designed and configured in stepped frequency continuous-wave mode to collect the vital sign returns of a subject from different angles and separating breathing and heartbeat rates. A vital signs model has been also proposed to study and simulate the system for any orientation.

A comprehensive study on the ability of FMCW radars to detect the vital signs for different body orientations has been described in [111]. The vital signs of five subjects were measured for different orientations with a 5.8 GHz radar and the results were collected and statistically analyzed by demonstrating the radar effectiveness in measuring both breathing and heartbeat for different orientations.

Passive radar is a recently developed architecture showing a great potential for healthcare applications. The main idea is to exploit signals already present in the environment as the radar EM source and obtain information based on the signal reflection from targets. This could enable radar-based passive sensing in different parts of the world, potentially reducing costs, energy consumption and EM emissions compared with conventional radar sensors. If the radar receives signals directly from the transmitter and indirectly from the target reflections, they can be differentiated and elaborated as in a single-mixer receiver. As an example, in [100], a 2.4 GHz passive radar has been designed based on a single-mixer input to keep the costs low. It has been successfully exploited to detect vital signs and hand-gestures.

Sleep monitoring is another widely investigated application of portable radars, due to the alleviated discomfort for the user compared to conventional contact sensors.

Although some examples have been reported in this section, sleep monitoring is a widely investigated field, worthy of a specific section but it involves a more articulated analysis related not only to vital signs but also movements, sleep stage, entering/leaving a bed, etc. detection [112]. It represents a good connection topic between this and the next section whereby a human activity analysis will be explored in detail.

A recent example can be found in [113], where the authors measured heartrate and sleep data with a radar chip integrated in a Google Nest Hub placed next to the bed. This radar location has been tested because current radar sensors are usually placed in front of the user's chest or back, making it more difficult to integrate them into portable devices. The radar chip is a very compact miniaturized Soli radar chip embedded in a portable device (Google Nest Hub), while measurements have been enhanced by advanced signal processing and machine learning techniques.

In [114], an FMCW radar was employed to monitor both breathing and heart rates during sleep. Ten subjects were tested in different positions, achieving a mean absolute error equal to 3.6% and 9.1% for the heart rate and breathing rate, respectively.

The scientific literature also shows valuable contributions where sleep monitoring radars are successfully exploited for concurrent vital sign and body movement detection [113].

Radar-based vital sign detection has also been widely investigated as an identification system for non-contact security [25], as an occupancy sensor for a great number of applications [19], [21], [116]–[118], and for many other applications already mentioned in the first section, such as searching for people after a natural disaster, continuous authentication and in-cabin detection [2], [14]–[18].

3.2 Human Activity Analysis

The purpose of detecting human activity without any contact is one of the more exciting radar application fields. As mentioned in Section 3.1, activity detection is often based on the analysis of the micro-Doppler behavior of the subject, according to what has been described in Section 3.2. However, multiple approaches exist based on mixing the information coming from different sensors or different radar information [92]. A lot of studies show machine learning techniques to enhance the classification performance of deep neural networks for motion classification [119].

The previous section described some examples of radar vital signs detection for sleep monitoring. However, radar features have been exploited during sleep not only for vital sign detection but also to monitor the changing sleep posture as a meaningful diagnostic assessment [120].

A widely investigated topic concerns counting the number of people within a room. This was also mentioned in the previous section and indeed often people counting involves vital sign detection. However, exploiting the micro-Doppler signature of humans performing activities can be very beneficial to enforcing people detection, thus making the counting task more reliable for different applications, such as those of surveillance, assisted living, and search and rescue. As an example, in [121] the authors apply machine learning techniques to perform human activity classification and people counting with high classification accuracy (above 97%), providing guidelines for machine learning modelling. Deep neural networks have also been used for motion classification of human activities in [122].

Concerning the topic of motion classification, recognizing the motion patterns of specific parts of the body from the radar returns is not an easy task. The reflecting signal can return from different flexible body articulations making many different motion patterns. As an example, limb trajectories have peculiar trajectories that can be exploited to understand if a specific subject is armed. This idea was investigated in [123], where the authors used the limb returns to discriminate between humans and animals and between armed and non-armed humans. This task was accomplished by implementing a modified Viterbi algorithm and Hough transform on the micro-Doppler signatures. The procedure was demonstrated experimentally. Joining range-Doppler and micro-Doppler signature, in [124] a potential active shooter was recognized. Tests were performed on a subject carrying a concealed rifle by applying an artificial neural network. An accuracy of 99.21% was demonstrated [124].

The swing of limbs was also studied in [125], using different micro-Doppler distributions to classify human activities. Moreover, a complex-value convolutional neural network was exploited for the automatic classification. A similar approach was followed in [126], but with a temporal range-Doppler PointNet-based method to detect anomaly behaviors.

With the ever-increasing average age of society, monitoring the health of elderly people is attracting huge attention both for daily safety and health monitoring, particularly for subjects who live alone. Exploiting a contactless technology such as the radar one can enable continuous subject monitoring without discomfort or privacy concerns typical of wearable sensors and cameras, respectively.

In [127], a preliminary study presenting an indoor monitoring system for human activity recognition was presented. Again, it exploits the micro-Doppler feature extracted and classified for different targets.

Gait recognition and analysis represents another interesting topic including many body parameters of interest for assessing the subject wellbeing. Frequently the gait analysis involves first the micro-Doppler signature measurement and afterward the classification. As an example, in [128], physical activity and walking periods were monitored with a mmWave radar in an indoor scenario while a deep learning network was trained by using the range-Doppler map and the performance of different deep learning models evaluated. In [129], a radar-based gait analysis was performed with the task to recognize disguised gaits without any prior information on the subject's particular condition.

How the micro-Doppler signature can be affected by the target direction of movement was investigated in [130]. Therefore, an angle-sensitive classifier has been proposed, based on a convolutional neural network.

With the aim of monitoring and preserving the health and safety of elderly people, fall detection represents an important topic, since a fall could lead to serious injuries, long hospitalization, or even death. In the case of fall, a prompt action from a third person is very beneficial to limit the damage, thus continuous and comfortable monitoring is again required.

In [131], an elderly people fall detection system is proposed. It is based on the micro-Doppler signature detection and convolutional neural networks and obtained 99.77% test accuracy. In [132], a 24 GHz CW radar was used to detect falls. The fall detection performance was enhanced also by implementing vital sign monitoring to reduce the number of false positives.

Due to the great number of contributions employing machine learning algorithms, in [133] the authors' purpose was to decrease the computational cost and complexity, thus the power consumption of deep neural network algorithms for fall detection. In detail, the proposed algorithm first detects a motion event based on a threshold method; thereafter a shallow neural network tries to take a preliminary decision on the possible occurrence of a fall event with very low computational cost; finally, a deep neural network confirms or not the fall event.

Among the different related topics of interest, posture recognition is another interesting field to assess and improve the user's health.

In [134], a 24 GHz SISO FMCW radar was employed for the concurrent localization and posture estimation of a human target achieving a 98% classification accuracy.

The human activity recognition was exploited also to analyze several physical activities. In [135], four different exercises, i.e. sit-ups, push-ups, squats, and jump rope were detected, classified and analyzed. As an example, the number of repetitions was counted for multiple subjects performing different exercises. Range-Doppler data and a micro-Doppler signature were employed to achieve the desired results with a 60 GHz FMCW radar.

Finally, it is worth recalling that the radar signal capability cannot be attenuated by many materials, thus enabling proper operations protected by a radome or through wall applications. To this aim two main challenges are the presence of large stationary clutter, such as a wall, which produces a strong echo that can overcome that of the human target of interest and the distortion of the human echo due to the attenuation, refraction, and multipath effects from the wall [136].

In [137], human motion was detected behind an opaque wall medium by exploiting a radar. Throughout an experimental test, the authors demonstrated the detection of different body parts first by using a high-resolution time-range map, then by high-pass filtering the detrimental effects of the wall present in the time range map and finally the most significant micro-Doppler features were highlighted by a range-max enhancement strategy.

As a final example, in [138], the authors explored the topic of radar through wall human sensing detection with a passive radar. By using a software-defined radio receiver, they detected the position of an indoor Wi-Fi access point, and exploited its signal to detect the Doppler/micro-Doppler of human motions inside a room.

3.3 Visually Impaired People Aids

According to the World Report on Vision edited by the World Health Organization, more than 2.2 billion people are affected by visual impairment [139].

As a matter of fact, the scientific community has put great effort into investigating new technologies to assist the mobility of people affected by serious blindness. Tools employing different technologies are available on the market and many electronic travel aids (ETAs) have been developed based on camera, laser, infrared and ultrasonic sensors.

Moreover, the radar community has also investigated new ways to help visually impaired people. The main advantages of radar technology compared to classic systems, including higher accuracy, fewer privacy concerns, more compact size for higher operating frequencies, and robustness against different ambient light and weather conditions, have been already detailed in the introduction.

A pioneering work, where the idea of employing an electromagnetic system to help blind people has been described, is reported in [140]. In detail, VNA and horn antennas working in the 1–6 GHz band have been used to send pulses towards a target. By exploiting an inverse Fourier transform the time delay between the system and the target has been measured and the target range computed. The system performance has been compared with an ultrasonic system showing better precision and the ability to detect more types of obstacles, e.g., a partially open door.

After this preliminary study, which was very basic, bulky and expensive, a lot of progress has been made.

The system described in [140] was improved in [141], by exploiting a portable VNA for testing the system with a blind volunteer under different test conditions, both indoor and outdoor with a maximum range of 3 m and range resolution of 12 cm.

Assisting the mobility of blind people is an interesting topic during physical activities such as running. Employing a lightweight and compact EM system can improve athletes' comfort during competitions or training. Indeed, currently blind runners are guided by another runner, thus limiting the possibility of free movement [142]. In [142], a system based on EM walls has been proposed, whereby the term "EM wall" refers to a mobile unit ahead of the runner that creates two electromagnetic waves radiated on the left and right sides. The runner is equipped with a receiving unit with vibro-tactile devices triggered when the athlete runs into an EM wall, thus being forced to run in the middle of the path.

In [143], an FMCW radar system was proposed to avoid the collision of visually impaired people with unexpected obstacles. The radar system, working at a central frequency of 80 GHz with a bandwidth of 4 GHz, i.e. 6 cm range resolution, was placed on the head of a visually impaired person. The radar was equipped with a stepper motor to obtain angular information and to map the environment.

In [144], a clinical study was described with 25 blind people each wearing a 24 GHz radar. The system was able to compensate small tilt errors or warn the user about an excessive system inclination. Haptic and acoustic feedback can be used to inform the person about the presence of obstacles.

Although these articles demonstrated that a radar system can be easily integrated into clothes or hats, the traditional white cane is still the most employed and accepted aid for visually impaired people due to the ease of use and cost-effectiveness [32]. However, it has limits related to the limited scanned area, not only concerning the range but also the obstacle elevation. Indeed, the classic cane oscillating movement is not suitable for detecting suspended obstacles, such as the branch of a tree.

To this aim, the authors of [34] asked for feedback from blind users, that, during interviews, suggested that one of the most important features of the ETA should be the positive reaction to a new device. Due to the self-confidence that the traditional white cane gives to the blind subjects, the authors' effort was directed towards the design of a radar-based detection system based on the integration with a traditional white cane. One of the main requisites of such a device is to be cost-effective and usable without long training periods. As a matter of fact, the known white-cane limitations should be overcome, e.g., in terms of detection of suspended obstacles at the height of the chest or the head and with a higher range. Moreover, providing information about the nature of the target, i.e. if it is a human or non-human target, would be very appreciated. Pushed by this requirement, the term "microwave cane" was born, referring to a classic white cane equipped with a radar-based detection system.

In [34], the preliminary requirements of a microwave white cane were drawn up. The system specifications were validated by means of simulations and the transmitting and receiving antenna were designed.

The problem of designing an antenna tailored for the specific applications was addressed also in [145–146]. To put the requirement of being lightweight first, microstrip technology was selected for the antenna design and the system advantages compared to ultrasonic and optical technology were described.

The first experimental example of microwave cane was reported in [35]. The effectiveness of a 24 GHz FMCW radar with a range resolution of 14 cm, and a maximum range of 5 m was demonstrated. Ad-hoc antennas have been designed with azimuthal and elevation patterns close to 12° and 40°, respectively. Whereas the narrow azimuthal pattern was designed to assign the horizontal scanning to the manual movement of the white cane, the wide elevation pattern also allows detecting obstacles at the height of the head. Many real obstacles used during daily life were detected and the antenna radiation pattern tested.

Many effective sensing solutions, regardless of the specific application, are based on the use of different technologies, each with their own specific pros and cons. Sensing obstacles for blind people is not an exception. In [147], [148], radar and RGB-D sensors were used together. While the FMCW radar detected the range and speed of obstacles, the RGB-D sensor separated multiple targets at the same range by exploiting color information and implementing a single deep neural network or object instance segmentation.

Exploring radar-based sensors for human navigation in reduced visibility and for blind people has been also the task of the INSPEX project, described in [149].

In [31], a radar-based white cane working at very high frequency, i.e. 122 GHz, was presented. Compared to all the previous examples, the novelty of this contribution relies on the peculiar capability of understanding if the obstacle is a human or non-human target. As a matter of fact, this last information can be gathered by detecting the vital signs, if present.

A simple scenario with only one walking person has been considered to test the effectiveness of the solution. As described in Section 3.2, the phase extraction should be performed for a specific range. However, if the target is moving, the phase should be measured for different ranges, thus making the extraction more challenging. A solution adopted by the scientific community relies first on a range alignment stage, whereby all the range bins corresponding to a specific target are recognized and aligned and a second stage whereby the phase can be straightforwardly extracted from the newly aligned range.

This reasoning should suggest that considering only a single moving target is a large simplification because all the targets (only one in this case) migrate to the same extent due to the constant relative motion among radar and targets. When multiple moving targets, each with its own trajectory, are present, a more complicated procedure is required to individually recognize and align each target trajectory.

This issue has been addressed in [33], where a mmWave radar has been used both to detect, locate and recognize the nature of multiple moving targets

By using the very high frequency of 122 GHz, a lightweight and compact system should be realized whereas the respiratory signal has been detected as the key element to separate humans and non-human targets throughout a novel range-alignment procedure.

After associating the information concerning the presence of vital signs to a specific target, the system could provide the information concerning the targets' nature to the user.

3.4 In-cabin Health Monitoring

Car transportation is increasingly important in our society. In addition to the great effort of industry and academics towards the development of even more smart sensors for autonomous driving and outdoor security, huge attention is directed towards in-cabin scenarios. To this purpose, in-vehicle sensing involves many applications from detecting the passengers' presence, to monitoring the driver's health or attention, or to try controlling the car without contact [150]. Among the different employed technologies, radar sensing has undergone great advancements.

Detecting the passengers' presence can be very valuable for radar to act as the key enabling technology of a contactless anti-abandon system to avoid children being left alone in a car. The simplest way to check for human presence with a radar system involves a range measurement which will give different results if a person is present on the seat [17].

However, measuring also the subjects' vital signs can lead to a more reliable human occupancy detection.

In addition to the continuous-wave radar detailed in this book, ultrawide-band (UWB), in particular impulse radio-ultra wide band (IR-UWB), radar has also played a crucial role in biomedical sensing and imaging. In [151], an IR-UWB has been exploited to count the number of people inside a car by recognizing the motion through exploiting a spatial-temporal-circulated gray level co-occurrence matrix and physiological features for people sensing. The dataset used in the work is made available by the authors.

A similar example was also reported in [152] whereby an IR-UWB radar was tested to detect dangerous situations, such as children or pets being left alone in a car. How the received waveform varies for different people arrangements inside the vehicle has been analyzed.

In [153], the radar occupancy sensing capabilities were specifically tailored for child presence detection and enforced by machine learning models.

Also, in [18], [154], a radar system equipped with neural networks to enhance the detection was proposed but using a MIMO radar.

In [15], a 60 GHz pulsed coherent radar was used to detect the presence of people and to measure the breathing rate. The effect of the car vibrations and random body motion was experimentally studied.

Many studies are aimed at monitoring the health of the passengers to increase their safety. Since in these cases, occupancy detection is not the only purpose, additional and more sophisticate techniques are required to accurately measure the parameters of interest. In [155] the capability of MIMO radars to recognize dysfunctional breathing patterns was investigated. The subject's torso was divided into six regions, and the amplitude and frequency of its motion were measured with the task to classify patterns associated with dysfunctional breathing.

In [156] the passengers' vital signs were measured by exploiting an UWB radar. The noisy effect of subtle body motion was taken into account to extract breathing and heartbeat.

An accurate analysis has been developed in [157] whereby the vital signs were measured to monitor the driver's health/consciousness in the presence of driver motion artifacts. The authors first removed the motion artifact with a compensation module followed by a periodicity check to highlight the vital signs; thereafter, they reconstruct breathing and heart rate and test the system in a real driving environment.

Monitoring the driver's attentiveness is indeed crucial for transportation safety; in [158] the authors' effort was directed towards the detection of head motion, exploited as an indicator of the driver's consciousness. Eight different driver head movements were measured with an FMCW radar mounted in the dashboard of a car and the related micro-Doppler signatures were classified after being converted into spectrogram images.

The purpose of [42] is similar but seven typical driving behaviors were investigated through time-Doppler spectrogram and range-Doppler trajectory with a special focus also on the performance dependence on the operating frequency, individual diversity, and AoA.

A smart research trend aimed at improving the driver and passengers comfort involves the radar-based gesture detection to control the car infotainment systems without any contact. Most of the contributions are based on the gesture detection by means of micro-Doppler analysis and classification algorithms [2], [3], [159]–[162]. Although most of the works use Doppler or FMCW radars, examples using IR-UWB radars also exist [163].

A great effort is also directed to hardware-level research aimed at improving the radar performance. Among the recent contributions, in [16] a 24 GHz radar was designed and tailored for in-cabin applications. In detail, an ad-hoc third-order intermodulation distortion cancellation structure to improve the linearity of the I/Q up-conversion mixer and a poly-phase filter with parasitic line inductance were integrated with the classic radar components by using a CMOS process.

In [164], with the purpose of providing a low-profile appearance system, textile antennas were developed to be embedded into the upholstery, using a new approach to measure vital signs by positioning the antennas in the side lumbar support of the car seat.

3.5 Assistive Devices for the Communication of People Affected by Neurodegenerative Pathologies

One of the most basic needs of humans concerns the capability to communicate with other individuals. However, some diseases make the subject unable to move and finally communicate, such as those related to neurodegenerative pathologies.

Probably as a consequence of the life expectancy increase, the World Health Organization estimated the number of people affected by dementia, one of the main symptoms associated with neurodegenerative pathologies, as approximately 55 million in 2019. This number is expected to increase to 139 million in 2050 [165].

Due to neuron death and progressive paralysis, the patients lose their ability to communicate by speaking. This scenario has pushed the scientific community towards the development of technological solutions to enable the subjects' communication. Since the head is the last affected part of the body, many communication systems are based on the detection of intentional head movements or eye blinking. After being detected, they can be interpreted as messages or commands.

Indeed, many camera-based systems use a display, where images depicting basic needs or simple words browse slowly and can be selected after an intentional movement detection. However, the low immunity to the light conditions and the raised privacy concerns are known limitations [166].

Other solutions make use of electrooculography and electroencephalography to directly detect brain signals; however, they require contact electrodes placed on the head of the subject [167], [168].

In order to overcome the limitations of currently employed technologies, radars represent a valid solution. Some works have demonstrated the possibility to detect intentional head motion or eye-blinking with radar [36], [37]. FMCW or Doppler radars detected intentional motion due to the range shift consequence of the head motion or eyelid small thickness. In order to separate intentional from physiological movement, the first ones need to be coded. As an example, two consecutive eye-blinks can be due to an intentional command as opposed to a single physiologic eye blink.

The abovementioned contributions were improved in [38], whereby a mmWave radar system was exploited to detect intentional eye blinking based on both the resulting displacement and micro-Doppler signature. Moreover, particular attention was paid to investigating real aspects affecting the measure such as the immunity to body movement and physiological activity. With the purpose of enhancing the system immunity to external movements, the radar was integrated on glass frames and the system effectiveness was evaluated by means of simulations and experimental tests. The best radar position on the glasses was investigated, and the 120 GHz operating frequency allows designing a compact and lightweight system.

In [169], a 77 GHz FMCW radar was used to detect eye blinking for different subjects and under particular conditions, e.g., at different distances, angles, and movements. To improve the detection an adaptive variational mode decomposition algorithm was exploited.

Mainly thought for the communication of patients affected by paralysis aphasia, a radar-based "eye-language" translator was proposed in [170]. The results have been used to design an eyeblink-to-voice translation.

4

Conclusion

This book describes the working principles of modern radars and provides an overview of recent applications of microwave and mmWave radars for healthcare applications.

In addition to the radar architectures discussed in this book, another significant innovation in biomedical radar over the past decade has been the development of injection-locking techniques. This advancement has led to the creation of a see-through-wall radar capable of detecting human vital signs using self-injection-locked radar technology [171]. Additionally, mutual-injection-locking has been employed to mitigate noise caused by random body movements [172].

Enabled by the rise of IoT technologies, the number of wireless sensors in daily life is rapidly increasing, with this trend expected to continue. As wireless device proliferation raises concerns about interference and spectrum usage, passive sensing has emerged as a key technique to maintain the synergistic relationship between sensing and communications. Consequently, some biomedical radar sensors now use passive radar architectures, leveraging existing radio signals such as Wi-Fi to monitor human behavior and vital signs [100]. Although passive sensing offers high-sensitivity motion detection, its performance heavily depends on the positioning of the third-party illuminator, target, and radar receiver. To address this, beamforming and advanced arrays have been developed to optimize detection angles in practical settings [173].

While advanced semiconductor technology and hardware architecture are crucial to the evolution of modern biomedical radar frontends, the development of biomedical radar sensors is increasingly supported by advanced signal processing and machine learning techniques. For example, short-time autocorrelation can retain individual heartbeats and correct anomalies caused by artifacts [174]. Wavelet transforms have been employed to minimize the impact of random body swaying on vital parameter estimation [175]. A differentiate-and-cross-multiply (DACM) algorithm was developed [176], leading to the potential for 1D imaging of human cardiac motion for clinical heart disease diagnosis. Empirical mode decomposition has been widely used to generate unique feature vectors for classifying human motions [177]. Deep learning has also been employed [178]. Readers interested in these topics are encouraged to explore articles published by the signal processing and radar systems communities.

The authors' aim is that this book represents a starting point for new students and researchers approaching this topic or a reference text for those who want to study specific sub-topics focused on the healthcare field of radar technologies. Ultimately, the authors hope to inspire joint efforts directed towards new innovative theories, technologies, techniques, and applications of modern biomedical radar.

References

[1] William L. Melvin, James A. Scheer, "*Principles of Modern Radar Vol. III: Radar Applications*" SciTech Publishing, Edison, NJ, 2014.

[2] E. Cardillo, L. Ferro, and C. Li, "Microwave and millimeter-wave radar circuits for the next generation contact-less in-cabin detection," Asia Pacific Microwave Conference (APMC), Yokohama, Japan., pp. 231-233, Nov. 2022. DOI: 10.23919/APMC55665.2022.9999764.

[3] E. Cardillo, L. Ferro, C. Li, and A. Caddemi, "Microwave radars for automotive in-cabin detection," Lecture Notes in Electrical Engineering, pp 75–80, Feb. 2023.

[4] Boric-Lubecke O., Lubecke V.M., Droitcour A.D., Park B.K., Singh A. Doppler *Radar Physiological Sensing*. Wiley; Hoboken, NJ, USA: 2016.

[5] Changzhi Li, Jenshan Lin Microwave Noncontact Motion Sensing and Analysis, Wiley; Hoboken, NJ, USA: 2014.

[6] D. Rodriguez and C. Li, "Sensitivity and Distortion Analysis of a 125-GHz Interferometry Radar for Submicrometer Motion Sensing Applications," IEEE Transactions on Microwave Theory and Techniques, vol. 67, no. 12, pp. 5384-5395, Dec. 2019, doi: 10.1109/TMTT.2019.2951142.

[7] F. Fioranelli, et al. "Radar sensing for healthcare," *Electron. Lett.,* 2019, 55, (9), pp. 1022–1024, doi: 10.1049/el.2019.2378.

[8] M. Alizadeh, G. Shaker, J. C. M. D. Almeida, P. P. Morita and S. Safavi-Naeini, "Remote Monitoring of Human Vital Signs Using mm-Wave FMCW Radar," in IEEE Access, vol. 7, pp. 54958-54968, 2019, doi: 10.1109/ACCESS.2019.2912956

[9] C. Li, V. M. Lubecke, O. Boric-Lubecke and J. Lin, "A Review on Recent Advances in Doppler Radar Sensors for Noncontact Healthcare Monitoring," in IEEE Transactions on Microwave Theory and Techniques, vol. 61, no. 5, pp. 2046-2060, May 2013, doi: 10.1109/TMTT.2013.2256924.

[10]   T. Lauteslager, M. Tommer, T. S. Lande and T. G. Constandinou, "Coherent UWB Radar-on-Chip for In-Body Measurement of Cardiovascular Dynamics," in IEEE Transactions on Biomedical Circuits and Systems, vol. 13, no. 5, pp. 814-824, Oct. 2019, doi: 10.1109/TBCAS.2019.2922775.

[11]   H. Zhao, X. Gu, H. Hong, Y. Li, X. Zhu and C. Li, "Non-contact Beat-to-beat Blood Pressure Measurement Using Continuous Wave Doppler Radar," 2018 IEEE/MTT-S International Microwave Symposium - IMS, Philadelphia, PA, USA, 2018, pp. 1413-1415, doi: 10.1109/MWSYM.2018.8439354.

[12]   D. Buxi, J. -M. Redouté and M. R. Yuce, "Blood Pressure Estimation Using Pulse Transit Time From Bioimpedance and Continuous Wave Radar," in IEEE Transactions on Biomedical Engineering, vol. 64, no. 4, pp. 917-927, April 2017, doi: 10.1109/TBME.2016.2582472.

[13]   E. Cardillo and A. Caddemi, "Radar Range-Breathing Separation for the Automatic Detection of Humans in Cluttered Environments," in IEEE Sensors Journal, vol. 21, no. 13, pp. 14043-14050, 1 July1, 2021, doi: 10.1109/JSEN.2020.3024961.

[14]   Islam, S.M.M.; Bori-Lubecke, O.; Zheng, Y.; Lubecke, V.M. Radar-Based Non-Contact Continuous Identity Authentication. Remote Sens. 2020, 12, 2279. https://doi.org/10.3390/rs12142279.

[15]   W. Li, R. J. Piechocki, K. Woodbridge, C. Tang and K. Chetty, "Passive WiFi Radar for Human Sensing Using a Stand-Alone Access Point," in IEEE Transactions on Geoscience and Remote Sensing, vol. 59, no. 3, pp. 1986-1998, March 2021, doi: 10.1109/TGRS.2020.3006387.

[16]   Lee, S.; Jeon, Y.; Park, G.; Myung, J.; Lee, S.; Lee, O.; Moon, H.; Nam, I. A 24-GHz RF Transmitter in 65-nm CMOS for In-Cabin Radar Applications. Electronics 2020, 9, 2005. https://doi.org/10.3390/electronics9122005 .

[17]   A. Caddemi and E. Cardillo, "Automotive Anti-Abandon Systems: a Millimeter-Wave Radar Sensor for the Detection of Child Presence," 2019 14th International Conference on Advanced Technologies, Systems and Services in Telecommunications (TELSIKS), Nis, Serbia, 2019, pp. 94-97, doi: 10.1109/TELSIKS46999.2019.9002193.

[18]   H. Abedi, M. Ma, J. He, J. Yu, A. Ansariyan and G. Shaker, "Deep Learning-Based In-Cabin Monitoring and Vehicle Safety System Using a 4-D Imaging Radar Sensor," in IEEE Sensors Journal, vol. 23, no. 11, pp. 11296-11307, 1 June1, 2023, doi: 10.1109/JSEN.2023.3270043.

[19]   E. Cardillo, C. Li and A. Caddemi, "Heating, Ventilation, and Air Conditioning Control by Range-Doppler and Micro-Doppler Radar Sensor," 2021 18th European Radar Conference (EuRAD), London, United Kingdom, 2022, pp. 21-24, doi: 10.23919/EuRAD50154.2022.9784461.

[20] A. Santra, R. V. Ulaganathan and T. Finke, "Short-Range Millimetric-Wave Radar System for Occupancy Sensing Application," in IEEE Sensors Letters, vol. 2, no. 3, pp. 1-4, Sept. 2018, Art no. 7000704, doi: 10.1109/LSENS.2018.2852263.

[21] E. Cardillo, C. Li, and A. Caddemi, "Embedded heating, ventilation, and air conditioning control systems: from traditional technologies towards radar advanced sensing," Review of Scientific Instruments, vol. 92, Issue 6, 061501, pp. 1-14, Jun. 2021. DOI: 10.1063/5.0044673.

[22] J. C. Lin, "Noninvasive microwave measurement of respiration," in Proceedings of the IEEE, vol. 63, no. 10, pp. 1530-1530, Oct. 1975, doi: 10.1109/PROC.1975.9992.

[23] A. D. Droitcour, O. Boric-Lubecke, V. M. Lubecke, J. Lin and G. T. A. Kovacs, "Range correlation and I/Q performance benefits in single-chip silicon Doppler radars for noncontact cardiopulmonary monitoring," in IEEE Transactions on Microwave Theory and Techniques, vol. 52, no. 3, pp. 838-848, March 2004, doi: 10.1109/TMTT.2004.823552.

[24] Victor C. Chen, "The Micro-Doppler Effect in Radar," Artech House 685 Canton St. Norwood, MA, 2019.

[25] C. Li, V. M. Lubecke, O. Boric-Lubecke and J. Lin, "Sensing of Life Activities at the Human-Microwave Frontier," in IEEE Journal of Microwaves, vol. 1, no. 1, pp. 66-78, Jan. 2021, doi: 10.1109/JMW.2020.3030722.

[26] Wang, F.-K.; Tang, M.-C.; Chiu, Y.-C.; Horng, T.-S. Gesture Sensing Using Retransmitted Wireless Communication Signals Based on Doppler Radar Technology. IEEE Trans. Microw. Theory Tech. 2015, 63, 4592–4602

[27] A. -K. Seifert, M. Grimmer and A. M. Zoubir, "Doppler Radar for the Extraction of Biomechanical Parameters in Gait Analysis," in IEEE Journal of Biomedical and Health Informatics, vol. 25, no. 2, pp. 547-558, Feb. 2021, doi: 10.1109/JBHI.2020.2994471.

[28] E. Cardillo, L. Ferro, and D. V. Q. Rodrigues, "Exploiting millimeter-wave radars to enable accurate gesture recognition for the metaverse environment," Lecture Notes in Electrical Engineering, pp. 110-115, Nov. 2023, doi: 10.1007/978-3-031-48711-8_13.

[29] Z. Peng, J. -M. Muñoz-Ferreras, R. Gómez-García and C. Li, "FMCW radar fall detection based on ISAR processing utilizing the properties of RCS, range, and Doppler," 2016 IEEE MTT-S International Microwave Symposium (IMS), San Francisco, CA, USA, 2016, pp. 1-3, doi: 10.1109/MWSYM.2016.7540121.

[30] E. Cardillo, C. Li, and A. Caddemi, "Radar-based monitoring of the worker activities by exploiting range-Doppler and micro-Doppler signatures," IEEE International Workshop on Metrology for

Industry 4.0 and IoT, Rome, Italy, pp. 412-416, Jun. 2021. DOI: 10.1109/MetroInd4.0IoT51437.2021.9488464.

[31] E. Cardillo, C. Li and A. Caddemi, "Empowering Blind People Mobility: A Millimeter-Wave Radar Cane," 2020 IEEE International Workshop on Metrology for Industry 4.0 & IoT, Roma, Italy, 2020, pp. 213-217, doi: 10.1109/MetroInd4.0IoT48571.2020.9138239.

[32] Cardillo, E.; Caddemi, A. Insight on Electronic Travel Aids for Visually Impaired People: A Review on the Electromagnetic Technology. Electronics 2019, 8, 1281. https://doi.org/10.3390/electronics8111281.

[33] E. Cardillo, C. Li and A. Caddemi, "Millimeter-Wave Radar Cane: A Blind People Aid With Moving Human Recognition Capabilities," in IEEE Journal of Electromagnetics, RF and Microwaves in Medicine and Biology, vol. 6, no. 2, pp. 204-211, June 2022, doi: 10.1109/JERM.2021.3117129.

[34] V. Di Mattia et al., "A feasibility study of a compact radar system for autonomous walking of blind people," 2016 IEEE 2nd International Forum on Research and Technologies for Society and Industry Leveraging a better tomorrow (RTSI), Bologna, Italy, 2016, pp. 1-5, doi: 10.1109/RTSI.2016.7740599.

[35] E. Cardillo et al., "An Electromagnetic Sensor Prototype to Assist Visually Impaired and Blind People in Autonomous Walking," in IEEE Sensors Journal, vol. 18, no. 6, pp. 2568-2576, 15 March15, 2018, doi: 10.1109/JSEN.2018.2795046.

[36] E. Cardillo, G. Sapienza, C. Li and A. Caddemi, "Head Motion and Eyes Blinking Detection: a mm-Wave Radar for Assisting People with Neurodegenerative Disorders," 2020 50th European Microwave Conference (EuMC), Utrecht, Netherlands, 2021, pp. 925-928, doi: 10.23919/EuMC48046.2021.9338116.

[37] E. Cardillo, G. Sapienza, L. Ferro, C. Li and A. Caddemi, "Radar Assistive System for People with Neurodegenerative Disorders Through Head Motion and Eyes Blinking Detection," 2023 IEEE/MTT-S International Microwave Symposium - IMS 2023, San Diego, CA, USA, 2023, pp. 979-982, doi: 10.1109/IMS37964.2023.10187979.

[38] E. Cardillo, L. Ferro, G. Sapienza and C. Li, "Reliable Eye-Blinking Detection With Millimeter-Wave Radar Glasses," in IEEE Transactions on Microwave Theory and Techniques, vol. 72, no. 1, pp. 771-779, Jan. 2024, doi: 10.1109/TMTT.2023.3329707.

[39] Z. Xiao et al. , "Human Eye Activity Monitoring Using Continuous Wave Doppler Radar: A Feasibility Study," in IEEE Transactions on Biomedical Circuits and Systems, vol. 18, no. 2, pp. 322-333, April 2024, doi: 10.1109/TBCAS.2023.3325547.

[40] J. Hu et al., "Real-Time Contactless Eye Blink Detection Using UWB Radar," in IEEE Transactions on Mobile Computing, vol. 23, no. 6, pp. 6606-6619, June 2024, doi: 10.1109/TMC.2023.3323280.

[41] H. N. Nguyen, S. Lee, T. Nguyen and Y. Kim, "One-shot learning-based driver's head movement identification using a millimetre-wave radar sensor", IET Radar Sonar Navigat., vol. 16, no. 5, pp. 825-836, May 2022.

[42] C. Ding et al., "Inattentive Driving Behavior Detection Based on Portable FMCW Radar," in IEEE Transactions on Microwave Theory and Techniques, vol. 67, no. 10, pp. 4031-4041, Oct. 2019, doi: 10.1109/TMTT.2019.2934413.

[43] D. V. Q. Rodrigues and C. Li, "Tracking Driver's Foot Movements Using mmWave FMCW Radar," 2024 IEEE Topical Conference on Wireless Sensors and Sensor Networks (WiSNeT), San Antonio, TX, USA, 2024, pp. 34-36, doi: 10.1109/WiSNeT59910.2024.10438602.

[44] Cardillo, E.; Caddemi, A. A Review on Biomedical MIMO Radars for Vital Sign Detection and Human Localization. Electronics 2020, 9, 1497. https://doi.org/10.3390/electronics9091497.

[45] T. Sakamoto, "Noncontact Measurement of Human Vital Signs during Sleep Using Low-power Millimeter-wave Ultrawideband MIMO Array Radar," 2019 IEEE MTT-S International Microwave Biomedical Conference (IMBioC), Nanjing, China, 2019, pp. 1-4, doi: 10.1109/IMBIOC.2019.8777864.

[46] B. R. Upadhyay, A. B. Baral and M. Torlak, "Vital Sign Detection via Angular and Range Measurements With mmWave MIMO Radars: Algorithms and Trials," in IEEE Access, vol. 10, pp. 106017-106032, 2022, doi: 10.1109/ACCESS.2022.3211527.

[47] Y. Li, C. Gu and J. Mao, "A Robust and Accurate FMCW MIMO Radar Vital Sign Monitoring Framework With 4-D Cardiac Beamformer and Heart-Rate Trace Carving Technique," in IEEE Transactions on Microwave Theory and Techniques, doi: 10.1109/TMTT.2024.3384288.

[48] Z. Peng and C. Li, "A Portable K-Band 3-D MIMO Radar With Nonuniformly Spaced Array for Short-Range Localization," in IEEE Transactions on Microwave Theory and Techniques, vol. 66, no. 11, pp. 5075-5086, Nov. 2018, doi: 10.1109/TMTT.2018.2869565.

[49] Lo Presti, D.; Carnevale, A.; D'Abbraccio, J.; Massari, L.; Massaroni, C.; Sabbadini, R.; Zaltieri, M.; Di Tocco, J.; Bravi, M.; Miccinilli, S.; et al. A multi-parametric wearable system to monitor neck movements and respiratory frequency of computer workers. Sensors 2020, 20, 536, doi:10.3390/s20020536.

[50] Massaroni, C.; Di Tocco, J.; Bravi, M.; Carnevale, A.; Presti, D.L.; Sabbadini, R.; Miccinilli, S.; Sterzi, S.; Formica, D.; Schena, E. Respiratory monitoring during physical activities with a multi-sensor smart

garment and related algorithms. IEEE Sens. J. 2020, 20, 2173–2180, doi:10.1109/JSEN.2019.2949608.

[51]   Rum, L.; Sciarra, T.; Balletti, N.; Lazich, A.; Bergamini, E. Validation of an Automatic Inertial Sensor-Based Methodology for Detailed Barbell Velocity Monitoring during Maximal Paralympic Bench Press. Sensors 2022, 22, 9904. https://doi.org/10.3390/s22249904.

[52]   A. Baldazzi, L. Molinaro, J. Taborri, F. Margheritini, St. Rossi, E. Bergamini, "Reliability of wearable sensors-based parameters for the assessment of knee stability" PLoS ONE 17(9): e0274817. doi.org/10.1371/journal.pone.0274817.

[53]   Dinh, T.; Nguyen, T.; Phan, H.-P.; Nguyen, N.-T.; Dao, D.V.; Bell, J. Stretchable respiration sensors: Advanced designs and multifunctional platforms for wearable physiological monitoring. Biosens. Bioelectron. 2020, 166, 112460, doi:10.1016/j.bios.2020.112460.

[54]   Webster J. Medical Instrumentation. Wiley; 2010.

[55]   Jubran A. Pulse oximetry. Tobin MJ (ed). Principles and Practice of Intensive Care Monitoring. New York: McGraw Hill, Inc.; 1998:261–287.

[56]   A. Leardini, L. Chiari, U. Della Croce, A. Cappozzo, Human movement analysis using stereophotogrammetry: Part 3. Soft tissue artifact assessment and compensation, Gait & Posture, Vol. 21, no. 2, pp. 212-225, 2005.

[57]   S. A. Shah and F. Fioranelli, "RF Sensing Technologies for Assisted Daily Living in Healthcare: A Comprehensive Review," in IEEE Aerospace and Electronic Systems Magazine, vol. 34, no. 11, pp. 26-44, 1 Nov. 2019, doi: 10.1109/MAES.2019.2933971.

[58]   Poh, M.-Z.; McDuff, D.J.; Picard, R.W. Advancements in noncontact, multiparameter physiological measurements using a webcam. IEEE Trans. Biomed. Eng. 2011, 58, 7–11, doi:10.1109/TBME.2010.2086456.

[59]   Sun, Y.; Thakor, N. Photoplethysmography revisited: From contact to noncontact, from point to imaging. IEEE Trans. Biomed. Eng. 2016, 63, 3, 463477, doi:10.1109/TBME.2015.2476337.

[60]   X. Ma, P. Wang, L. Chen, F. Zhang and D. Zhang, "Mitigation of UWB Radar Self-Motion for mm-Scale Vibration Detection," 2022 IEEE MTT-S International Wireless Symposium (IWS), Harbin, China, 2022, pp. 1-3, doi: 10.1109/IWS55252.2022.9978061.

[61]   S. Bennett, T. N. El Harake, R. Goubran and F. Knoefel, "Adaptive Eulerian Video Processing of Thermal Video: An Experimental Analysis," in IEEE Transactions on Instrumentation and Measurement, vol. 66, no. 10, pp. 2516-2524, Oct. 2017, doi: 10.1109/TIM.2017.2684518.

[62]   Indie Semiconductors (formerly Silicon Radar) TRA_120_002_V1.1 Datasheet, Frankfurt, Germany, 2023.

[63] M. Garbey, N. Sun, A. Merla and I. Pavlidis, "Contact-Free Measurement of Cardiac Pulse Based on the Analysis of Thermal Imagery," in IEEE Transactions on Biomedical Engineering, vol. 54, no. 8, pp. 1418-1426, Aug. 2007, doi: 10.1109/TBME.2007.891930.

[64] Infineon BGT24MTR11 datasheet, Rev. 3.1, 2014-03-25.

[65] Texas Instruments IWR6843, IWR6443 Single-Chip 60- to 64-GHz mmWave Sensor, 2021.

[66] E. Cardillo, C. Li and A. Caddemi, "Vital Sign Detection and Radar Self-Motion Cancellation Through Clutter Identification," in IEEE Transactions on Microwave Theory and Techniques, vol. 69, no. 3, pp. 1932-1942, March 2021, doi: 10.1109/TMTT.2021.3049514.

[67] Ferro, L.; Li, C.; Scandurra, G.; Ciofi, C.; Cardillo, E. Beneficial Effects of Self-Motion for the Continuous Phase Analysis of Ac-Coupled Doppler Radars. Electronics 2024, 13, 772. https://doi.org/10.3390/electronics13040772.

[68] Jingxuan Chen, Yajie Wu, Bo Zhang, Shisheng Guo and Guolong Cui (2024), "A Lightweight Remote Gesture Recognition System with Body-motion Suppression and Foreground Segmentation Using FMCW Radar", APSIPA Transactions on Signal and Information Processing: Vol. 13: No. 4, e304. http://dx.doi.org/10.1561/116.0000006.

[69] Yang, S.; Liang, X.; Dang, X.; Jiang, N.; Cao, J.; Zeng, Z.; Li, Y. Random Body Movement Removal Using Adaptive Motion Artifact Filtering in mmWave Radar-Based Neonatal Heartbeat Sensing. Electronics 2024, 13, 1471. https://doi.org/10.3390/electronics13081471.

[70] T. Sakamoto, D. Sanematsu, I. Iwata, T. Minami and M. Myowa, "Radar-Based Respiratory Measurement of a Rhesus Monkey by Suppressing Nonperiodic Body Motion Components," in IEEE Sensors Letters, vol. 7, no. 10, pp. 1-4, Oct. 2023, Art no. 7005204, doi: 10.1109/LSENS.2023.3311672.

[71] S. Hazra et al., "Robust Radar-Based Vital Sensing With Adaptive Sinc Filtering and Random Body Motion Rejections," in IEEE Sensors Letters, vol. 7, no. 5, pp. 1-4, May 2023, Art no. 7001604, doi: 10.1109/LSENS.2023.3266237.

[72] K. Han and S. Hong, "MIMO Differential Radar Using Null Point Beams for Vital Sign Detection in the Presence of Body Motions," 2022 19th European Radar Conference (EuRAD), Milan, Italy, 2022, pp. 177-180, doi: 10.23919/EuRAD54643.2022.9924869.

[73] K. Han, B. N. Nibret and S. Hong, "Body Motion Artifact Cancellation Technique for Cough Detection Using FMCW Radar," in IEEE Microwave and Wireless Technology Letters, vol. 33, no. 1, pp. 106-109, Jan. 2023, doi: 10.1109/LMWC.2022.3198174.

[74] D.M. Pozar, "Microwave Engineering," 4th Edition. Wiley, New York, 2011.

[75]   Mark A. Richards, James A. Scheer, William A. Holm *"Principles of Modern Radar Vol. I: Basic Principles,"* SciTech Publishing, Edison, NJ, 2011.

[76]   Long, M.W., "Radar Clutter," Tutorial presented at the 2006 IEEE Radar Conference, Verona, NY, April 2006.

[77]   A. Caddemi and E. Cardillo, "A study on dynamic threshold for the crosstalk reduction in frequency-modulated radars," 2017 Computing and Electromagnetics International Workshop (CEM), Barcelona, Spain, 2017, pp. 29-30, doi: 10.1109/CEM.2017.7991871.

[78]   E. Cardillo and A. Caddemi, "A novel approach for crosstalk minimization in frequency modulated continuous wave radars," Electronics Letters, Vol. 53, Issue 20, pp. 1379-1381, Sept. 2017. DOI: 10.1049/el.2017.2800.

[79]   S. Yuan, F. Fioranelli and A. Yarovoy, "An adaptive threshold-based unambiguous robust Doppler beam sharpening algorithm for forward-looking MIMO Radar," 2023 20th European Radar Conference (EuRAD), Berlin, Germany, 2023, pp. 65-68, doi: 10.23919/EuRAD58043.2023.10288626.

[80]   Y. Su, T. Cheng and Z. He, "Joint Resource and Detection Threshold Optimization for Maneuvering Targets Tracking in Colocated MIMO Radar Network," in IEEE Transactions on Aerospace and Electronic Systems, vol. 59, no. 5, pp. 5900-5914, Oct. 2023, doi: 10.1109/TAES.2023.3267336.

[81]   X. Guan, L. -h. Zhong, D. -h. Hu and C. -b. Ding, "Threshold determination for dynamic programming based track-before-detect in passive bistatic radar," 2014 12th International Conference on Signal Processing (ICSP), Hangzhou, China, 2014, pp. 1938-1943, doi: 10.1109/ICOSP.2014.7015331.

[82]   A. D. Droitcour, O. Boric-Lubecke, V. M. Lubecke and Jenshan Lin, "0.25 /spl mu/m CMOS and BiCMOS single-chip direct-conversion Doppler radars for remote sensing of vital signs," 2002 IEEE International Solid-State Circuits Conference. Digest of Technical Papers (Cat. No.02CH37315), San Francisco, CA, USA, 2002, pp. 348-349 vol.1, doi: 10.1109/ISSCC.2002.993075.

[83]   A. D. Droitcour, O. Boric-Lubecke, V. M. Lubecke, J. Lin and G. T. A. Kovacs, "Range correlation effect on ISM band I/Q CMOS radar for non-contact vital signs sensing," IEEE MTT-S International Microwave Symposium Digest, 2003, Philadelphia, PA, USA, 2003, pp. 1945-1948 vol.3, doi: 10.1109/MWSYM.2003.1210539.

[84]   L. Ferro, G. Scandurra, C. Li and E. Cardillo, "Robust Doppler Displacement Measurement Resolving the Uncertainty During Target Stationary Moment," 2024 IEEE Topical Conference on Wireless Sensors and

Sensor Networks (WiSNeT), San Antonio, TX, USA, 2024, pp. 57-60, doi: 10.1109/WiSNeT59910.2024.10438649.

[85]   Gill, T.P., The Doppler Effect, Logos Press, London, 1965.

[86]   Temes, C.L., "Relativistic Consideration of Doppler Shift," IRE Transactions on Aeronautical and Navigational Electronics, p. 37, 1959.

[87]   Faruk Uysal "Comparison of range migration correction algorithms for range Doppler processing," Journal of Applied Remote Sensing 11(3), 036023. https://doi.org/10.1117/1.JRS.11.036023.

[88]   W. -C. Su, Y. -C. Lai, T. -S. Horng and R. E. Arif, "Time-Division Multiplexing MIMO Radar System With Self-Injection-Locking for Image Hotspot-Based Monitoring of Multiple Human Vital Signs," in IEEE Transactions on Microwave Theory and Techniques, vol. 72, no. 3, pp. 1943-1952, March 2024, doi: 10.1109/TMTT.2023.3308159.

[89]   H. Reggad, X. Jiang, X. Wu, R. Amirtharajah, D. Matthews and X. Liu, "A Single-Chip Single-Antenna Radar for Remote Vital Sign Monitoring," in IEEE Transactions on Microwave Theory and Techniques, vol. 71, no. 10, pp. 4519-4532, Oct. 2023, doi: 10.1109/TMTT.2023.3267554.

[90]   B. -K. Park, O. Boric-Lubecke and V. M. Lubecke, "Arctangent Demodulation With DC Offset Compensation in Quadrature Doppler Radar Receiver Systems," in IEEE Transactions on Microwave Theory and Techniques, vol. 55, no. 5, pp. 1073-1079, May 2007, doi: 10.1109/TMTT.2007.895653.

[91]   F. Fioranelli, J. Le Kernec and S. A. Shah, "Radar for Health Care: Recognizing Human Activities and Monitoring Vital Signs," in IEEE Potentials, vol. 38, no. 4, pp. 16-23, July-Aug. 2019, doi: 10.1109/MPOT.2019.2906977.

[92]   E. Cardillo and A. Caddemi, "Feasibility Study to Preserve the Health of an Industry 4.0 Worker: a Radar System for Monitoring the Sitting-Time," 2019 II Workshop on Metrology for Industry 4.0 and IoT (MetroInd4.0&IoT), Naples, Italy, 2019, pp. 254-258, doi: 10.1109/METROI4.2019.8792905.

[93]   Z. Peng, C. Li and F. Uysal, "Modern Radar for Automotive Applications," SciTech Publishing, Edison, NJ, 2022.

[94]   G. Richard Curry, Radar System Performance Modeling, Second Edition , Artech, 2004.

[95]   E. Cardillo, R. Cananzi, L. Ferro, C. Li, P. Vita and P. Vita, "Detection of Space Debris through Compact X Band FMCW Radar," 2023 16th International Conference on Advanced Technologies, Systems and Services in Telecommunications (TELSIKS), Nis, Serbia, 2023, pp. 126-129, doi: 10.1109/TELSIKS57806.2023.10316106.

[96]   Zhizang, C.; Gopal, K.G.; Yiqiang, Y. Introduction to Direction-of-Arrival Estimation, 1st ed.; Artech House: London, UK, 2010.

[97]   Balanis, C.A. Antenna Theory: Analysis and Design, 3rd ed.; Wiley: New York, NY, USA, 2005.

[98]   Li, J.; Stoica, P. MIMO Radar Signal Processing, 1st ed.; Wiley: New York, NY, USA, 2009.

[99]   S. Dong et al., "A Review on Recent Advancements of Biomedical Radar for Clinical Applications," in IEEE Open Journal of Engineering in Medicine and Biology, doi: 10.1109/OJEMB.2024.3401105.

[100]   D. Tang, V. G. R. Varela, D. V. Q. Rodrigues, D. Rodriguez and C. Li, "A Wi-Fi Frequency Band Passive Biomedical Doppler Radar Sensor," in IEEE Transactions on Microwave Theory and Techniques, vol. 71, no. 1, pp. 93-101, Jan. 2023, doi: 10.1109/TMTT.2022.3193408.

[101]   S. M. M. Islam, L. C. Lubecke, C. Grado and V. M. Lubecke, "An Adaptive Filter Technique for Platform Motion Compensation in Unmanned Aerial Vehicle Based Remote Life Sensing Radar," 2020 50th European Microwave Conference (EuMC), Utrecht, Netherlands, 2021, pp. 937-940, doi: 10.23919/EuMC48046.2021.9338011.

[102]   Z. Li, T. Jin, Y. Dai and Y. Song, "Motion-Robust Contactless Heartbeat Sensing Using 4-D Imaging Radar," in IEEE Transactions on Instrumentation and Measurement, vol. 72, pp. 1-10, 2023, Art no. 4011110, doi: 10.1109/TIM.2023.3312477.

[103]   X. Jiang, X. Gao, X. Wu, Q. J. Gu and X. Liu, "Automatic RF Leakage Cancellation for Improved Remote Vital Sign Detection Using a Low-IF Dual-PLL Radar System," in IEEE Transactions on Microwave Theory and Techniques, vol. 71, no. 6, pp. 2664-2679, June 2023, doi: 10.1109/TMTT.2022.3231658.

[104]   W. Xia, Y. Li and S. Dong, "Radar-Based High-Accuracy Cardiac Activity Sensing," in IEEE Transactions on Instrumentation and Measurement, vol. 70, pp. 1-13, 2021, Art no. 4003213, doi: 10.1109/TIM.2021.3050827.

[105]   W. A. Ahmad et al., "Multimode W-Band and D-Band MIMO Scalable Radar Platform," in IEEE Transactions on Microwave Theory and Techniques, vol. 69, no. 1, pp. 1036-1047, Jan. 2021, doi: 10.1109/TMTT.2020.3038532.

[106]   Y. Xiong, Z. Peng, C. Gu, S. Li, D. Wang and W. Zhang, "Differential Enhancement Method for Robust and Accurate Heart Rate Monitoring via Microwave Vital Sign Sensing," in IEEE Transactions on Instrumentation and Measurement, vol. 69, no. 9, pp. 7108-7118, Sept. 2020, doi: 10.1109/TIM.2020.2978347.

[107]   C. Feng et al., "Multitarget Vital Signs Measurement With Chest Motion Imaging Based on MIMO Radar," in IEEE Transactions on Microwave Theory and Techniques, vol. 69, no. 11, pp. 4735-4747, Nov. 2021, doi: 10.1109/TMTT.2021.3076239.

[108] S. M. M. Islam, O. Boric-Lubecke and V. M. Lubekce, "Concurrent Respiration Monitoring of Multiple Subjects by Phase-Comparison Monopulse Radar Using Independent Component Analysis (ICA) With JADE Algorithm and Direction of Arrival (DOA)," in IEEE Access, vol. 8, pp. 73558-73569, 2020, doi: 10.1109/ACCESS.2020.2988038.

[109] M. Mercuri et al., "2-D Localization, Angular Separation and Vital Signs Monitoring Using a SISO FMCW Radar for Smart Long-Term Health Monitoring Environments," in IEEE Internet of Things Journal, vol. 8, no. 14, pp. 11065-11077, 15 July15, 2021, doi: 10.1109/JIOT.2021.3051580.

[110] W. Ren et al., "Vital Sign Detection in Any Orientation Using a Distributed Radar Network via Modified Independent Component Analysis," in IEEE Transactions on Microwave Theory and Techniques, vol. 69, no. 11, pp. 4774-4790, Nov. 2021, doi: 10.1109/TMTT.2021.3101655.

[111] Sacco, G.; Piuzzi, E.; Pittella, E.; Pisa, S. An FMCW Radar for Localization and Vital Signs Measurement for Different Chest Orientations. Sensors 2020, 20, 3489. https://doi.org/10.3390/s20123489.

[112] H. Hong et al., "Microwave Sensing and Sleep: Noncontact Sleep-Monitoring Technology With Microwave Biomedical Radar," in IEEE Microwave Magazine, vol. 20, no. 8, pp. 18-29, Aug. 2019, doi: 10.1109/MMM.2019.2915469.

[113] Xu, L., Lien, J., Li, H. et al. Soli-enabled noncontact heart rate detection for sleep and meditation tracking. Sci Rep 13, 18008 (2023). https://doi.org/10.1038/s41598-023-44714-2.

[114] Turppa, E.; Kortelainen, J.M.; Antropov, O.; Kiuru, T. Vital Sign Monitoring Using FMCW Radar in Various Sleeping Scenarios. Sensors 2020, 20, 6505. https://doi.org/10.3390/s20226505.

[115] B. Yu et al., "WiFi-Sleep: Sleep Stage Monitoring Using Commodity Wi-Fi Devices," in IEEE Internet of Things Journal, vol. 8, no. 18, pp. 13900-13913, 15 Sept.15, 2021, doi: 10.1109/JIOT.2021.3068798.

[116] S. M. M. Islam, A. Droitcour, E. Yavari, V. M. Lubecke, O. Boric-Lubecke, "Building occupancy estimation using microwave Doppler radar and wavelet transform," Building and Environment, Vol. 236, 110233, May 2023, doi: 10.1016/j.buildenv.2023.110233.

[117] A. Lazaro, M. Lazaro, R. Villarino and D. Girbau, "Seat-Occupancy Detection System and Breathing Rate Monitoring Based on a Low-Cost mm-Wave Radar at 60 GHz," in IEEE Access, vol. 9, pp. 115403-115414, 2021, doi: 10.1109/ACCESS.2021.3105390.

[118] T. Grebner, P. Schoeder, V. Janoudi and C. Waldschmidt, "Radar-Based Mapping of the Environment: Occupancy Grid-Map Versus SAR," in IEEE Microwave and Wireless Components Letters, vol. 32, no. 3, pp. 253-256, March 2022, doi: 10.1109/LMWC.2022.3145661.

[119] B. Erol, S. Z. Gurbuz and M. G. Amin, "Motion Classification Using Kinematically Sifted ACGAN-Synthesized Radar Micro-Doppler Signatures," in IEEE Transactions on Aerospace and Electronic Systems, vol. 56, no. 4, pp. 3197-3213, Aug. 2020, doi: 10.1109/TAES.2020.2969579.

[120] J. E. Kiriazi, S. M. M. Islam, O. Bori-Lubecke and V. M. Lubecke, "Sleep Posture Recognition With a Dual-Frequency Cardiopulmonary Doppler Radar," in IEEE Access, vol. 9, pp. 36181-36194, 2021, doi: 10.1109/ACCESS.2021.3062385.

[121] F. Luo, S. Poslad and E. Bodanese, "Human Activity Detection and Coarse Localization Outdoors Using Micro-Doppler Signatures," in IEEE Sensors Journal, vol. 19, no. 18, pp. 8079-8094, 15 Sept.15, 2019, doi: 10.1109/JSEN.2019.2917375.

[122] X. Li, Y. He, F. Fioranelli and X. Jing, "Semisupervised Human Activity Recognition With Radar Micro-Doppler Signatures," in IEEE Transactions on Geoscience and Remote Sensing, vol. 60, pp. 1-12, 2022, Art no. 5103112, doi: 10.1109/TGRS.2021.3090106.

[123] Y. Ding, R. Liu, Y. She, B. Jin and Y. Peng, "Micro-Doppler Trajectory Estimation of Human Movers by Viterbi–Hough Joint Algorithm," in IEEE Transactions on Geoscience and Remote Sensing, vol. 60, pp. 1-11, 2022, Art no. 5113111, doi: 10.1109/TGRS.2022.3171208.

[124] Y. Li, Z. Peng, R. Pal and C. Li, "Potential Active Shooter Detection Based on Radar Micro-Doppler and Range-Doppler Analysis Using Artificial Neural Network," in IEEE Sensors Journal, vol. 19, no. 3, pp. 1052-1063, 1 Feb.1, 2019, doi: 10.1109/JSEN.2018.2879223.

[125] X. Yao, X. Shi and F. Zhou, "Human Activities Classification Based on Complex-Value Convolutional Neural Network," in IEEE Sensors Journal, vol. 20, no. 13, pp. 7169-7180, 1 July1, 2020, doi: 10.1109/JSEN.2020.2967054.

[126] M. Li, T. Chen and H. Du, "Human Behavior Recognition Using Range-Velocity-Time Points," in IEEE Access, vol. 8, pp. 37914-37925, 2020, doi: 10.1109/ACCESS.2020.2975676.

[127] Z. Wang, A. Ren, Q. Zhang, A. Zahid and Q. H. Abbasi, "Recognition of Approximate Motions of Human Based on Micro-Doppler Features," in IEEE Sensors Journal, vol. 23, no. 11, pp. 12388-12397, 1 June1, 2023, doi: 10.1109/JSEN.2023.3267820.

[128] H. Abedi, A. Ansariyan, P. P. Morita, A. Wong, J. Boger and G. Shaker, "AI-Powered Noncontact In-Home Gait Monitoring and Activity Recognition System Based on mm-Wave FMCW Radar and Cloud Computing," in IEEE Internet of Things Journal, vol. 10, no. 11, pp. 9465-9481, 1 June1, 2023, doi: 10.1109/JIOT.2023.3235268.

[129] Y. Yang, D. Zhao, X. Yang, B. Li, X. Wang and Y. Lang, "Open-Scenario-Oriented Human Gait Recognition Using Radar Micro-Doppler Signatures," in IEEE Transactions on Aerospace and Electronic Systems, doi: 10.1109/TAES.2024.3403077.

[130] Y. Yang, C. Hou, Y. Lang, T. Sakamoto, Y. He and W. Xiang, "Omni-directional Motion Classification With Monostatic Radar System Using Micro-Doppler Signatures," in IEEE Transactions on Geoscience and Remote Sensing, vol. 58, no. 5, pp. 3574-3587, May 2020, doi: 10.1109/TGRS.2019.2958178.

[131] F. J. Abdu, Y. Zhang and Z. Deng, "Activity Classification Based on Feature Fusion of FMCW Radar Human Motion Micro-Doppler Signatures," in IEEE Sensors Journal, vol. 22, no. 9, pp. 8648-8662, 1 May1, 2022, doi: 10.1109/JSEN.2022.3156762.

[132] K. Hanifi and M. E. Karsligil, "Elderly Fall Detection With Vital Signs Monitoring Using CW Doppler Radar," in IEEE Sensors Journal, vol. 21, no. 15, pp. 16969-16978, 1 Aug.1, 2021, doi: 10.1109/JSEN.2021.3079835.

[133] J. Lu and W. -B. Ye, "Design of a Multistage Radar-Based Human Fall Detection System," in IEEE Sensors Journal, vol. 22, no. 13, pp. 13177-13187, 1 July1, 2022, doi: 10.1109/JSEN.2022.3177173.

[134] S. Yang and Y. Kim, "Single 24-GHz FMCW Radar-Based Indoor Device-Free Human Localization and Posture Sensing With CNN," in IEEE Sensors Journal, vol. 23, no. 3, pp. 3059-3068, 1 Feb.1, 2023, doi: 10.1109/JSEN.2022.3227025.

[135] V. G. Rizzi Varela, D. V. Q. Rodrigues, L. Zeng and C. Li, "Multitarget Physical Activities Monitoring and Classification Using a V-Band FMCW Radar," in IEEE Transactions on Instrumentation and Measurement, vol. 72, pp. 1-10, 2023, Art no. 8500910, doi: 10.1109/TIM.2022.3227998.

[136] T. Pardhu, V. Kumar, P. Kumar and N. Deevi, "Advancements in UWB-Based Human Motion Detection Through Wall: A Comprehensive Analysis," in IEEE Access, vol. 12, pp. 89818-89835, 2024, doi: 10.1109/ACCESS.2024.3397465.

[137] Q. An et al., "Range-Max Enhanced Ultrawideband Micro-Doppler Signatures of Behind-the-Wall Indoor Human Motions," in IEEE Transactions on Geoscience and Remote Sensing, vol. 60, pp. 1-19, 2022, Art no. 5107219, doi: 10.1109/TGRS.2021.3122138.

[138] H. Sun, L. G. Chia and S. G. Razul, "Through-Wall Human Sensing With WiFi Passive Radar," in IEEE Transactions on Aerospace and Electronic Systems, vol. 57, no. 4, pp. 2135-2148, Aug. 2021, doi: 10.1109/TAES.2021.3069767.

[139] 2019 World Report on Vision, The World Health Organization website. [Online]. Available: http://www.who.int/. Last accessed: July 2024.

[140] L. Scalise et al., "Experimental Investigation of Electromagnetic Obstacle Detection for Visually Impaired Users: A Comparison With Ultrasonic Sensing," in IEEE Transactions on Instrumentation and Measurement, vol. 61, no. 11, pp. 3047-3057, Nov. 2012, doi: 10.1109/TIM.2012.2202169.

[141] V. Di Mattia et al., "An electromagnetic device for autonomous mobility of visually impaired people," 2014 44th European Microwave Conference, Rome, Italy, 2014, pp. 472-475, doi: 10.1109/EuMC.2014.6986473.

[142] Pieralisi, M.; Petrini, V.; Di Mattia, V.; Manfredi, G.; De Leo, A.; Scalise, L.; Russo, P.; Cerri, G. Design and realization of an electromagnetic guiding system for blind running athletes. Sensors 2015, 15, 16466–1648316466-16483. https://doi.org/10.3390/s150716466.

[143] P. Kwiatkowski, T. Jaeschke, D. Starke, L. Piotrowsky, H. Deis and N. Pohl, "A concept study for a radar-based navigation device with sector scan antenna for visually impaired people," 2017 First IEEE MTT-S International Microwave Bio Conference (IMBIOC), Gothenburg, Sweden, 2017, pp. 1-4, doi: 10.1109/IMBIOC.2017.7965796.

[144] Kiuru, T.; Metso, M.; Utriainen, M.; Metsavainio, K.; Jauhonen, H.M.; Rajala, R.; Savenius, R.; Strom, M.; Jylha, T.N.; Juntunen, R.; et al. Assistive device for orientation and mobility of the visually impaired based on millimeter wave radar technology—Clinical investigation results. Cogent Eng. 2018, 5, 1–12, doi: 10.1080/23311916.2018.1450322.

[145] Pisa, S.; Pittella, E.; Piuzzi, E. Serial patch array antenna for an FMCW radar housed in a white cane. Int. J. Antennas Propag. 2016, 2016, 9458609, doi: 10.1155/2016/9458609.

[146] A. De Leo, P. Russo and G. Cerri, "Electronic Travel Aid for Visually Impaired People: Design and Experimental of a Special Antenna," 2021 IEEE International Symposium on Medical Measurements and Applications (MeMeA), Lausanne, Switzerland, 2021, pp. 1-6, doi: 10.1109/MeMeA52024.2021.9478670.

[147] Long, N.; Wang, K.; Cheng, R.; Hu, W.; Yang, K. Unifying obstacle detection, recognition, and fusion based on millimeter wave radar and RGB–depth sensors for the visually impaired. Rev. Sci. Instrum. 2019, 90, 044102, doi: 10.1063/1.5093279.

[148] Long, N.; Wang, K.; Cheng, R.; Yang, K.; Hu, W.; Bai, J. Assisting the visually impaired: Multitarget warning through millimeter wave radar and RGB–depth sensors. J. Electron. Imag. 2019, 28, 013028, doi: 0.1117/1.JEI.28.1.013028.

[149] Lesecq, S.; Foucault, J.; Birot, F.; De Chaumont, H.; Jackson, C.; Correvon, M.; Heck, P.; Banach, R.; Di Matteo, A.; Di Palma, V.; et al. INSPEX: Design and integration of a portable/wearable smart spatial exploration system. In Proceedings of the 2017 Design, Automation &

Test in Europe Conference & Exhibition, Lausanne, Switzerland, 27–31 Mar. 2017 pp. 746-751, doi: 10.23919/DATE.2017.7927089.

[150] A. Gharamohammadi, A. Khajepour and G. Shaker, "In-Vehicle Monitoring by Radar: A Review," in IEEE Sensors Journal, vol. 23, no. 21, pp. 25650-25672, 1 Nov.1, 2023, doi: 10.1109/JSEN.2023.3316449.

[151] X. Yang, Y. Ding, X. Zhang and L. Zhang, "Spatial-Temporal-Circulated GLCM and Physiological Features for In-Vehicle People Sensing Based on IR-UWB Radar," in IEEE Transactions on Instrumentation and Measurement, vol. 71, pp. 1-13, 2022, Art no. 8502113, doi: 10.1109/TIM.2022.3165808.

[152] S. Lim, S. Lee, J. Jung and S. -C. Kim, "Detection and Localization of People Inside Vehicle Using Impulse Radio Ultra-Wideband Radar Sensor," in IEEE Sensors Journal, vol. 20, no. 7, pp. 3892-3901, 1 April1, 2020, doi: 10.1109/JSEN.2019.2961107.

[153] K. Sato, S. Wandale, K. Ichige, K. Kimura and R. Sugiura, "Millimeter-Wave Radar-based Vehicle In-Cabin Occupancy Detection Using Explainable Machine Learning," in IEEE Sensors Journal, doi: 10.1109/JSEN.2024.3413775.

[154] Z. Xiao, K. Ye and G. Cui, "PointNet-Transformer Fusion Network for In-Cabin Occupancy Monitoring With mm-Wave Radar," in IEEE Sensors Journal, vol. 24, no. 4, pp. 5370-5382, 15 Feb.15, 2024, doi: 10.1109/JSEN.2023.3347893.

[155] M. -J. López, C. Palacios Arias, J. Romeu and L. Jofre-Roca, "In-Cabin MIMO Radar System for Human Dysfunctional Breathing Detection," in IEEE Sensors Journal, vol. 22, no. 24, pp. 23906-23914, 15 Dec.15, 2022, doi: 10.1109/JSEN.2022.3221052.

[156] K. -K. Shyu, L. -J. Chiu, P. -L. Lee and L. -H. Lee, "UWB Simultaneous Breathing and Heart Rate Detections in Driving Scenario Using Multi-Feature Alignment Two-Layer EEMD Method," in *IEEE Sensors Journal*, vol. 20, no. 17, pp. 10251-10266, 1 Sept.1, 2020, doi: 10.1109/JSEN.2020.2992687.

[157] F. Wang, X. Zeng, C. Wu, B. Wang and K. J. R. Liu, "Driver Vital Signs Monitoring Using Millimeter Wave Radio," in *IEEE Internet of Things Journal*, vol. 9, no. 13, pp. 11283-11298, 1 July1, 2022, doi: 10.1109/JIOT.2021.3128548

[158] D. G. Bresnahan and Y. Li, "Classification of Driver Head Motions Using a mm-Wave FMCW Radar and Deep Convolutional Neural Network," in *IEEE Access*, vol. 9, pp. 100472-100479, 2021, doi: 10.1109/ACCESS.2021.3096465.

[159] X. zhang, Q. Wu and D. Zhao, "Dynamic Hand Gesture Recognition Using FMCW Radar Sensor for Driving Assistance," *2018 10th International Conference on Wireless Communications and*

Signal Processing (WCSP), Hangzhou, China, 2018, pp. 1-6, doi: 10.1109/WCSP.2018.8555642.

[160]   K. A. Smith, C. Csech, D. Murdoch and G. Shaker, "Gesture Recognition Using mm-Wave Sensor for Human-Car Interface," in *IEEE Sensors Letters*, vol. 2, no. 2, pp. 1-4, June 2018, Art no. 3500904, doi: 10.1109/LSENS.2018.2810093.

[161]   P. Molchanov, S. Gupta, K. Kim and K. Pulli, "Multi-sensor system for driver's hand-gesture recognition," *2015 11th IEEE International Conference and Workshops on Automatic Face and Gesture Recognition (FG)*, Ljubljana, Slovenia, 2015, pp. 1-8, doi: 10.1109/FG.2015.7163132.

[162]   P. Molchanov, S. Gupta, K. Kim and K. Pulli, "Short-range FMCW monopulse radar for hand-gesture sensing," *2015 IEEE Radar Conference (RadarCon)*, Arlington, VA, USA, 2015, pp. 1491-1496, doi: 10.1109/RADAR.2015.7131232.

[163]   Khan, F.; Leem, S.K.; Cho, S.H. Hand-Based Gesture Recognition for Vehicular Applications Using IR-UWB Radar. *Sensors* **2017**, *17*, 833. https://doi.org/10.3390/s17040833.

[164]   C. Gouveia, C. Loss, P. Pinho, J. Vieira and D. Albuquerque, "Low-Profile Textile Antenna for Bioradar Integration Into Car Seat Upholstery: Wireless acquisition of vital signs while on the road," in *IEEE Antennas and Propagation Magazine*, vol. 66, no. 1, pp. 22-33, Feb. 2024, doi: 10.1109/MAP.2023.3254484.

[165]   [2022 World Report on Vision Webpage on The World Health Organization. [Online]. Available: https://www.who.int/publications/i/item/978924151657 0].

[166]   L. Ren, L. Kong, F. Foroughian, H. Wang, P. Theilmann, and A. E. Fathy, "Comparison study of noncontact vital signs detection using a Doppler stepped-frequency continuous-wave radar and camera-based imaging photoplethysmography," IEEE Trans. Microw. Theory Techn., vol. 65, no. 9, pp. 3519–3529, Sep. 2017, doi: 10.1109/TMTT.2017.2658567.

[167]   T. Wibble, T. Pansell, S. Grillner, and J. Pérez-Fernández, "Conserved subcortical processing in visuo-vestibular gaze control," Nature Commun., vol. 13, no. 1, p. 4699, Aug. 2022.

[168]   W. Deng, J. Huang, S. Kong, Y. Zhan, J. Lv, and Y. Cui, "Pupil trajectory tracing from video-oculography with a new definition of pupil location," Biomed. Signal Process. Control, vol. 79, Jan. 2023, Art. no. 104196.

[169]   Xinze Zhang, Walid Brahim, Mingyang Fan, Jianhua Ma, Muxin Ma, and Alex Qi. 2023. Radar-Based Eyeblink Detection Under Various Conditions. In Proceedings of the 2023 12th International Conference on Software and Computer Applications (ICSCA '23). Association for Computing Machinery, New York, NY, USA, 177–183. https://doi.org/10.114 5/3587828.3587855.

[170] Qi, F. et al. (2024). Radar Translator: Contactless Eyeblink Detection Assisting Basic Daily Intension Voice for the Paralyzed Aphasia Patient Using Bio-Radar. In: Wang, G., Yao, D., Gu, Z., Peng, Y., Tong, S., Liu, C. (eds) 12th Asian-Pacific Conference on Medical and Biological Engineering. APCMBE 2023. IFMBE Proceedings, vol 104. Springer, Cham. https://doi.org/10.1007/978-3-031-51485-2_39.

[171] F. Wang, T. Horng, K. Peng, J. Jau, J. Li and C. Chen, "Detection of concealed individuals based on their vital signs by using a see-through-wall imaging system with a self-injection-locked radar," IEEE Trans. Microwave Theory Tech., vol. 61, pp. 696-704, 2012.

[172] F. Wang, T. Horng, K. Peng, J. Jau, J. Li and C. Chen, "Single-antenna Doppler radars using self and mutual injection locking for vital sign detection with random body movement cancellation," IEEE Trans. Microwave Theory Tech., vol. 59, pp. 3577-3587, 2011.

[173] A. B. Carman and C. Li, "A Digital Beamforming Fast-Start Passive Radar for Indoor Motion Detection and Angle Estimation," in IEEE Transactions on Microwave Theory and Techniques, doi: 10.1109/TMTT.2024.3391062.

[174] N.T.P. Nguyen, P. Lyu, M.H. Lin, C. Chang and S. Chang, "A short-time autocorrelation method for noncontact detection of heart rate variability using CW doppler radar," in 2019 IEEE MTT-S International Microwave Biomedical Conference (IMBioC), pp. 1-4, 2019.

[175] T.K.V. Dai, Y. Yu, P. Theilmann, A.E. Fathy and O. Kilic, "Remote Vital Sign Monitoring with Reduced Random Body Swaying Motion Using Heartbeat Template and Wavelet Transform Based on Constellation Diagrams," IEEE J.Electromagn.RF Microw.Med.Biol, pp. 1-8, 2022.

[176] J. Wang, X. Wang, L. Chen, J. Huangfu, C. Li and L. Ran, "Noncontact distance and amplitude-independent vibration measurement based on an extended DACM algorithm," IEEE Transactions on Instrumentation and Measurement, vol. 63, pp. 145-153, 2013.

[177] D.P. Fairchild and R.M. Narayanan, "Classification of human motions using empirical mode decomposition of human micro-Doppler signatures," IET Radar, Sonar & Navigation, vol. 8, pp. 425-434, 2014.

[178] B. Jokanovi and M. Amin, "Fall detection using deep learning in range-Doppler radars," IEEE Trans.Aerospace Electron.Syst., vol. 54, pp. 180-189, 2017.

Index

Printed in the United States
by Baker & Taylor Publisher Services